Wild Life

Wild Life

Volume 1

Pat Neal

Writers Club Press
San Jose New York Lincoln Shanghai

Wild Life

Writers Club Press
an imprint of iUniverse.com, Inc.

For information address:
iUniverse.com, Inc.
5220 S 16th, Ste. 200
Lincoln, NE 68512
www.iuniverse.com

Cover design by Lily Neal.

ISBN: 0-595-16812-4

Printed in the United States of America

To my mom, Claire Neal.

Epigraph

While many names of places, people, and events may have been changed to protect the guilty, any resemblance to actual events, locales, or people living or dead is unfortunate. This is a work of non-fiction. All of these stories are true, and if they're not, they should be.

Contents

List of Illustrations

Foreword

These columns were originally published in a newspaper that prefers not to be identified for reasons that will become obvious once you read them.

A Guide's Toughest Clients

My name is Pat and I am a fishaholic. I think I have always had a fishing problem. Looking back I guess fishing has affected my schooling, jobs, family, and stuff. I just never knew how much until the day I couldn't fish. Fishing is a disease. The more you fish, the more you have to fish until the only way to fish as much as you have to is to be a millionaire or a fishing guide. I am a fishing guide.

I help people with fishing problems. Ever bought prawns with your food stamps and used them for bait? Ever tied a fly with spotted owl feathers? You could have a fishing problem. It's not something you have to be ashamed of any more. Half the people who buy a steelhead license in Washington State don't catch one. They are skunked. Skunking can lead to depression or even…golf.

River guiding is a competitive game, where death threats are the sincerest form of flattery. Disputes are settled with gaff hooks at ten paces. Where people hate you for being happy. Happiness you see is a lot like fishing. There is only so much on earth.

I became a guide to help people find happiness. That, with the money, power, and revenge, I call "sharing my love of nature." But every sport has its dark side. Every major league has its trash-talking upstarts who ruin it for everyone.

People have always asked who was my toughest client. Was it the drunk who said he was going to kill a fish or kill the guide? Heck no, I like drunks. Sometimes they over pay. I liked his attitude. Together we worked out some of his self-esteem issues that liberated him from his "victim" mentality. Sure I skunked him. He just wanted to cut me up a little. I'm a people person.

Was it the fly fisherman who got in the boat and bragged that his Spey rod and gold reel were worth more than everything I owned? Simple fact. He left that fancy rigging on top of his car when he drove back to the big city.

I saw the rod later on a plunking bar downriver, jammed in the sand, braced up by a forked beaver stick. There was a knuckle buster salmon reel clamped on to the handle and a bell taped on to the tip. Some kid had rigged the noble Spey to catch a first steelhead, which would have been one more than the fly fisherman ever caught with it.

With the exuberance of youth she related hocking the reel and cutting five feet off the tip of the rod to get just the right action. Tears welled as the simple kindness of that fly fisherman passed our angling heritage to another generation.

These are my friends. I'll fish them any time. The toughest clients had to be the gang in the front of the boat last weekend chanting, "Boring, boring, boring." Then there was that endless question, "Are we there yet? There yet? There?" and I was the one doing the asking.

They were tough. Instant gratification wasn't fast enough for them. They wanted a fish in the boat every minute and a half. They wanted to net it, club it, gut it and poke its eyes with a stick. Catch and release? That's for people who don't know how to run a smokehouse.

I'd like to toss them overboard into the rapids, except they'd think it was fun and tell me to do it again, and again, and again. Like I need another chore.

These are my kids. You're supposed to take your kids fishing, right? What happens if you take your kids fishing too much? Read on.

A real guide can tell by the signs if there are rough seas ahead. It all started with a tough launch. The bowline slipped. The boat slid off the rock ledge and shot into the middle of the glacier river like a runaway fish duck.

The kids thought this was funny. For them the fishing trip was over. The boat and all my gear were about to be sucked under a logjam. The kids knew I was over insured. They had the money already spent on a trip to Disneyland.

As much as I hated to disappoint them, I ran out and got the boat. Guides often wade the river, I explained, to see how deep the water is so we'll know what lures to run.

By the time we got on the water it was after eleven, which is when the beautiful people eat lunch. I wanted to get a line in. The crew was hostile.

Like an idiot I said I always got a bite in this hole. It's not like I'd lie to my own kids, but self-delusion is an important skill in this business. I was just thinking out loud.

So we got three plugs out, which is a gold-lettered invitation to a tangle. But the crew knew the drill and we made it halfway through the hole without a rigging fit or a bite. Then the chanting started. "Boring, boring, boring."

Fish on! The center rod went down and that shut them up. A big hen steelhead thrashed on the surface and peeled line downstream full speed ahead. The kid knew what to do. She's been fishing since before she was born. Lulu is her name. That's her sister's name too. I wanted to name them Ray and Bob, after the Raybobber, the greatest steelhead lure ever made, but then, they were girls and I'm really bad with names.

Lulu likes the heavy gear. A gaff hook would have been her choice, but I forced her to use the light tackle, eighteen-pound line on a hotshot magnum rod. The fish jumped, the crew screamed. Lulu gave not slack. We drifted downstream to a riffle. Lulu reefed until the fish was right on the surface. Her big sister grabbed the net and made a stab. The fish was in the bag!

It was a convulsive dinosaur of a trout she had to hoist into the boat immediately. "Let it do its fighting on the deck," Lulu said. You can take her word for that. She's been fishing hard for twelve years, if you count the trips before she was born. It's all time on the water, we figure.

Lulu hoisted the prize. The plug flew out of the fish's mouth. The fish slid through a hole in the net. Everyone started yelling at once, at me.

We'd been out crabbing the day before in Dungeness Bay. It's a tiny inlet barely six miles long, maybe two miles wide. Nose into that puddle with a sixteen-foot drift boat and you're bound to hit something. A guide should have known better.

Our landing net had snagged a channel marker some fool had pounded in the mud. Must have torn a hole.

Lulu demanded lunch but I was sitting in front of the cooler rowing so she couldn't get at the fizzies and junk. We were fishing. I told her the other guides probably cut that hole in the net just because they were jealous, so lash it up quick.

The gear was working the best hole on the Olympic Peninsula, Washington State, or the universe. It was a beautiful sunny day without another person in sight. A bear walked out of the woods and shuffled down the shore.

"Shoot it, Pa!" Lulu pointed. I tried to explain how a pack of inbred do-gooders had passed some fool law that made varmint hunting illegal. I could go to jail for a year.

"At least we could get a bear rug." Lulu sniffed. That hurt, like waking up with a skunk on your chest. Do you move real fast or hold real still? Thank the old fisherman the starboard rod went off. It was a good hit but the line went slack. There was a collective moan, then more yelling, at me. The fish hit again!

It rolled on the surface and tore a torpedo wake downriver. The line ripped a hole in the river. We followed through rapids into a tight cornered logjam hole. Thoughts of a broken oar, a flipped boat passed quickly. We're fishing. You expect to lose a little gear sometimes.

We skidded over a pile of rocks and plowed into the logjam. Wrecking a boat is like a lot of things. You don't really know how fast you're going until you stop. It was okay.

We shoved off the logjam. The kid reeled in hard. The steelhead was still on, though deep beneath a logjam tangled in an underwater rain-forest. "Looks like you caught us a big timber trout," Lulu smirked. Maybe a real guide could have dived in the river and caught that fish in his teeth. That's what Lulu said. Lucky for me I was on vacation.

I told the kid to give the fish some slack. This is a big deal on the river. Slack line means a lost fish. But we were snagged up already, so what the heck. Maybe it would just come loose on its own. Stranger things have happened.

I tried the pithy observation that just seeing a trophy steelhead in the 90's in America is a special life-affirming experience. This line works on ignorant tourists, but I shouldn't have tried it on Lulu. They want to touch the fish. I know.

Steelhead are the most beautiful animal that swims. They are a living rainbow trimmed silver and blue. They swim thousands of miles through dangerous seas where everything wants to eat them. They fight upstream through rapids far into the mountains, to a secret home in the hidden gravel.

To hold a steelhead after a hard fight is to touch the mysteries of migration, life and death in the wild. Wild life. Watch the least quiver of the tail push the fish upstream, further, home. I know what it means to touch the fish. They have touched me.

Lulu hit the free spool. We drifted downstream. I told her to bust it off but she just wouldn't listen. Then the fish was free! It held to the bottom of the logjam hole, looking for another snag. Steelhead are good at that. It's like they have every rock and stick in the river memorized, just in case.

Once they get tangled up they can break heavy line just by shaking their head, something they can't do against the action of a good rod with Lulu reefing for all she's worth.

Then the fish was at the surface. "It's a twenty pounder!" I said, and for once I really meant it. Lulu went for the net. Her sister belayed that. "He's just an old spawner." Lulu said, "Look at that slack belly. He's headed downstream tail first." I had to agree as I looked at the fish. Lulu was looking at me.

October 14, 1998

How a Guide Gets Clients

"You've got to be out of your mind if you think that thing will catch a fish," the client said as we floated down the Hoh River in the dark. She acted like she'd never seen a pink plastic worm before.

Fortunately that myth about the customer always being right has not trickled down to river guiding. People come here from all over the world to fish. With the lack of sleep, rough grub, pounding rides, the weather, and other hazing rituals we guides employ, the client has little chance of being right about anything. This leaves the poor person at the mercy of the guide, who is often in the same boat.

We're after some of the rarest animals that swim, the steelhead and the king salmon. They are as unpredictable as the rivers they run. With twelve feet of rain a year, conditions are subject to change without notice. Here today, gone later today. Anything you catch can be the fish of a lifetime. But sometimes hooking up takes longer than finding the remote control to your television.

Which brings me to the point of this article. How does a guide get clients anyway?

Most guys have trouble finding a buddy to sit out in a howling rainstorm getting skunked. Try charging money for the same experience. What with the government nailing on one law after another until no one can even figure out if the season is open, forget about finding clients.

I became a guide because I wanted to help people with fishing problems. But people have to admit they have a problem before they can be helped. This can be a painful experience, like not catching a fish for a couple of years. So I don't go looking for clients. That would be like a lawyer chasing car wrecks or a doctor with a trampoline dealership. The clients look for me.

Take Miss Wonderful for instance. She called shortly after the bars closed, which is a good time to leave Sequim if you want to be on the Hoh River before daylight in summer. Ordinarily we guides don't pick up clients in bars at night. It's against our strict code of ethics. Something to do with the old days in Port Townsend or Aberdeen, where a night on the town could include a drop through a trap door to the tide flats below and it's, "Have you been to sea, Billy?" Shanghaied. Now, I only go to bars for medical advice.

Then they said we couldn't pick up escaped prisoners from the vast gulag system on the upper Clearwater River either. Those boys were good business. It was a short trip with a good tip across the old Hoh River and no whining about getting skunked either. Why should an escaped honoree care about fish anyway? They were headed north to the bright lights of Dead Dog Flats, Sappho, or Hecklesville, fleshpots we call "the other Washington."

Shortly after I dropped some informative brochures by the old honor camp, another government do-gooder strung a jag of razor wire around the place. Sort of took the honor out of the honor camp if you ask me. It's just another example of big brother's jackboot breaking the back of the entrepreneurial spirit that made our country so cool.

So it's no good looking for clients. The clients have to find me. How they do it doesn't matter. I only want to help. I explained this to Miss Wonderful. "Open the bail. Let me hear your fish call," I gently instructed.

She had a darn good fish call. And do you know what? She caught a steelhead. I have the best job in the world.

June 27, 1998

Getting Lost in the Wilderness

Lately someone asked how you can avoid getting lost in the wilderness. That's a new one. I thought getting lost was the whole point of going into the wilderness in the first place. If you're worried about it, there are many nice and tidy places you can go and always know exactly where you are. But there are fewer and fewer places these days to go and get lost.

Getting lost is an important skill for a guide, like reading people's minds by looking at the back of their heads, or the ability to smell money, or the lack of.

I can get lost anywhere. Even in one of those fancy National Park Service campgrounds. It's a nature experience to trail the fabled comfort station while the clients enjoy the complimentary "guide special" breakfast of salmon jerky, hardtack, and special sauce.

It can take until nine or ten in the morning for them to get out of the bathroom once we find it, which is when I like to go fishing anyway. The people are better adjusted to their new surroundings. They've discarded the flowery naiveté that previously deluded them into thinking they would have a nice day. They left that attitude somewhere back with breakfast. They are ready to experience the thrill of nature on a new level, to meet the challenge of wilderness survival.

A day on the river can be a spiritual experience. I've seen people so overcome with the joy of suckling the breast of nature, they will drop to their knees and kiss the ground the minute they finally see their rental car again.

They've spent the day asking questions at the central core of the human experience. Where are we now? Where are we going? When will this trip be over?

I've spent the day telling them I have no idea. With the rain we've had this winter, the rivers flood about once a week. This shuffles the river channel and makes each day special. The fact is, no one could possibly know what's around the next bend of the river unless you fly over it first. This is not a T.V. show. It's a lot more hassle and you can't change the channel when it starts raining.

So we spend many days completely lost on a river I've floated hundreds of times. Where there was a fishing hole last week, there is now an island. Where there was an island, there is a channel that braids into three, choose one. If it shrinks into a trickle and dribbles down a knothole leaving you stranded hundreds of yards from the river, you took the wrong branch. It's okay. The float trip is now a backpacking trip. Backpacking is one of the best ways to get lost. The secret is to get lost in the right place. You'll need a good fishing hole with a sandy beach, cooking rocks, and a good stack of driftwood.

The main thing you have to remember is to not tell anyone where you're going or when you're coming back or you'll get the search and rescue do-gooders to come looking for you. They won't be happy. They'll find your secret fishing hole. You've been warned.

I always like to have a big map along when I'm lost. If you can keep it dry, a map can be good for starting a fire. As for a compass, there are many nice ones on the market. They are all worthless because you have to take a compass reading before you get lost. And if you're not lost, why bother with reading a compass? Toughen up.

We are fishing. We are explorers of new water. You could be the first or the last to fish it, depending on the weather. So far the weather has been bad.

It tests the resolve of the summer-run fisherman to fish in the teeth of a winter storm, to watch a tree older than Columbus totter in the gale and crash into the river, causing questions to be asked. If a tree falls in the forest, will anyone hear it?

I don't know about that, but if a tree falls in the river and smuts my boat, it will be just another funny story at Bob's Tackle in Forks, and everyone will hear about it. Nothing cheers up the locals like a dead guide.

All fishermen are brothers. It is a kinship beyond words. I have loaned the wheel bearings out of my own boat trailer to fellow anglers who needed them more than I did. It was the least I could do for friends I haven't met yet. Maybe they were lost. Maybe we are all lost. Some of us just don't care.

March 3, 1999

Endangered Species

It's the time of year I wait for. The Vernal Equinox. That's spring for you hillbillies. I know it's spring because I killed the first mosquito. It was the size of a swallow. I saw a swallow. The skunk cabbage is blooming. That bright thing came out almost two days in a row. This panicked the locals, sending rain gear sales into the gutter, pulling the plug on vinyl futures. That bright thing can be rough on the fishing. Ever had a tough time driving into a sunset? A steelhead has the same trouble seeing the gear when the sun's shining down the river.

So we launch early on a clear day, but it's worth it. It's fun to float the Sol Duc River in the dark. There's a heavy mist on the water. The mossy trees assume a ghostly shape. The still pools reflect the stars and the changing color of the sky.

It's a quiet time to engage in cheerful banter with my fellow anglers. How did they get here? How did they find me? Do they believe in Santa Claus?

We hear the gentle murmur of white water and pitch through a couple of chutes, where you too can assume a ghostly shape. The good people express apprehension. I explain how this is my first day on the river. I just escaped from a mental institution. How they'd never look for me out here because you'd have to be nuts to be a fishing guide in the first place.

Relax. I'm way over-insured. When I get a couple of people in the front of the boat I feel like a million bucks. Keep your hands and feet inside the boat. Listen to the voices. Not the ones telling you to kill a rock star, or get a job, that's crazy talk. Listen to the river voices, the gentle murmur of the current caressing the stones. Don't hear the river voices? Have another beer. It's noon somewhere.

On a good day the eagles fly upriver at daylight. These days they are packing branches to spruce up the nest. They're riding the thermals, mating in broad daylight. It's a disgusting spectacle I'd hoped to spare the children. Sometimes they're better off watching T.V.

The eagles spook packs of mergansers dressed in gaudy mating plumage. They whistle by grunting like pigs. They are pigs, sucking up the great downstream migration of young salmonids starting just now.

The dipper, or water ouzel is another obnoxious river pest. They're small black birds the size of starlings that can walk underwater. They stand on rocks and bob up and down like idiots. They never shut up. Throw in a rabble of wrens, siskins, herons, and woodpeckers and you can forget about having that quiet day on the river. The bird problem seems to be getting worse.

Here and there are tasteful homes perched above the water's edge whose inhabitants seem unaware of our passing in the dark. People who live in glass houses shouldn't.

As the day brightens, I thread a sand shrimp on a hook. This doesn't hurt him a bit. He wants to go fishing too. Give it a toss. Wait for a bite. And wait and wait.

Feel something wiggle? Reel it in. Maybe you'll land it, maybe you won't. Whatever happens, feel free to blame the guide. It's a service I provide all my clients. Blame insurance.

Some days I can guarantee you'll see a majestic Roosevelt elk. That's if one happens to be washed up in a logjam. They're wild animals. They move around. People are often disappointed upon viewing the flood victim elk, until I explain it's their shore lunch.

Sometimes people will ask if I will be making lunch. I have to explain how we're not married; they can't talk to me that way. This can set the tone for the day. People will start asking all kinds of crazy questions like, "Why would they log right to the edge of the river and leave a strip of timber to blow down and rot?"

I have to explain how a logger doesn't get up in the morning and decide where to run his saw. The ribbon stringers and the eggheads lay out the timber sales in compliance with strict environmental rules.

What may appear to the casual observer as an ugly, stupid, waste of our nation's forest resource is really a stream-side buffer designed to save our salmon.

The big deal over streamside buffers is a failed attempt to legislate common sense. It's another reason we deserve the endangered species act. Whatever that means.

April 14, 1999

Guide Showdown

Daylight on the river. Waiting for the stars to fade. Kick the fire together. Add a few more sparks to the universe. I try to do my part. It's the best time of day. It's all downhill from here.

It's the last day of the season. A time for quiet reflection on the passing year. Of one flood after another for months on end. The salmon and steelhead have been spawning in the high water, up in the brush at times. Now it is spring and the rivers are low. The eggs are high and dry. It's an environmental catastrophe you don't hear anything about. So forget that quiet reflection stuff. Life's too short. It's time to go fishing.

Then there is a sound. At first the low moaning of a pregnant buffalo, then louder until it is the choking growl of a jake brake on Fairholm hill.

You hear a lot of strange noises on the river. As a guide it's my job to pretend to know what they are. Cougar? Bear? Sasquatch? No, it is just another worn out fisherman, tired from a late night of re-arranging his tackle box.

Maybe you do your camping in some fancy RV or a sissy mountain tent. A real steelhead fisherman can sleep anywhere in just rain gear with a pair of wet hip boots for a pillow.

The cold light of day finds our angler becoming one with nature. It is a clear morning without frost, which means the mosquitoes are awake.

Lucky for him there are no mosquitoes on the Olympic Peninsula. What we have are some really small bloodsucking birds.

His eyes appear to be swelled shut and he hasn't even woke up yet. I think I should wake him up before he runs out of blood, or I won't get paid. But then I know from hard experience the minute they wake up they start wanting stuff. Daylight really is the best time of day.

There comes a sound of distant drums. Fertility ritual of a pagan tribe lost in the rainforest these many years? Sorry, that's another column. Someone is dragging a drift boat through the mud.

They have parked in a puddle then waddled through the crime scene tape to warm by the fire. They are offered the remains of last night's leftover "shore lunch" and tidings of plenteous luck. There are questions. Some people can't handle the truth. There is the shaking of heads. They float into the vapors. More boats launch. Everyone is in such a hurry. I am in a hurry, to be last.

That's my secret. I'm only telling you because there are no secrets on the river. Just when you think there are, along comes another guide who launches behind you. I know there is going to be a showdown, with a couple of boatloads of witnesses.

"How many fish have you caught?" he caught me off guard with the opening salvo. I tried to recover but then he added, "In your life."

"I caught ten yesterday. Twelve of them were over twenty pounds." I knew it was lame but I was tired and out of shape.

"That's as good as it gets," the guide replied, "for you."

"Look, a bald eagle." I pointed above, into the great beyond. It's an old trick, banned in all respectable bird-watching circles. Soon as they turn around to look, you tell them it's gone. I bird-watch with a pretty rough crowd. It was just another dirty trick that doesn't work on our man.

"I'll tell my clients," he said without bothering to look. "They pay me five dollars every time I see an eagle."

My vision blurred at that. He moved in closer to introduce himself saying, "I've been guiding here forever."

"How come you're still alive?" I asked

"I never said I was. Maybe you're fishing in hell."

"You go first," I pointed downriver. He pushed off, floated a little ways then anchored up in my favorite secret hole.

There was silence around the campfire. Maybe I should have tried a sheep joke. It was too late. It takes a tough hide to be a guide and it hurts to grow one. There are no free lessons in the school of hard knocks.

The end of fishing is the beginning of bird watching. They're migrating and mating. They don't care if you watch. Life is change. What was a worthless pack of smolt-sucking fish ducks are now the rare and colorful merganser. The rare and colorful harlequin ducks are thicker than rats at the dump. It's just a question of marketing. Look, there's a gyrfalcon!

April 28, 1999

A Guide's Toughest Hike

Lately someone asked what the toughest hike I've ever been on was. That was a hard one. If you think about it there are no real tough hikes left, only dumb hikers. I've been on some dumb hikes. Into the Elkhorn up the Elwha in the dark, back in the old days when a flashlight was considered extra weight for sissies. Or packing those log truck inner tubes over the Chilkoot Pass to make a raft to float down the Yukon when I could have mailed the works. Or packing that big bull elk out of…never mind they were all hikes to Dumbsville.

The toughest hikes I've been on have been down the halls of Children's Hospital to get Lulu a CAT scan. It was seven years ago today she had the cancer. Everybody knows about cancer. It's the worst thing in the world. A little kid with cancer is even worse than that. It's like a war with an invisible stalker. "It's like a fishing trip," one surgeon told me, "You never know what you're going to find." They don't sugar coat stuff much at Children's.

So she went in for the bonus plan treatment, surgery, radiation, and nine months of prophylactic chemotherapy. Personally I never liked hospitals and now I know why. The cure can be as bad as the disease.

Things started out bad, then got steadily worse. Any time the doctors stop talking medicine and start talking about God, you know you have a problem. And we were the lucky ones.

That's the great thing about Children's. You don't have to look far to find someone worse off than you are. I remember wondering how people did it, the care and feeding of a bald kid. A nice nurse said, "Get some help." So we did. At first I was embarrassed that we needed help. Later our motto became, "Spare us the sympathy, send cash." This community gave us thousands of dollars. The Cancer Society gave us ferry tickets and hotel rooms. Ronald McDonald House put us up for nothing.

As the good doctor said, "Might as well be having fun." But fun was mighty tough to come by. So the Make-a-Wish Foundation sent her to the funnest place on earth, Disneyland. She was in kind of a hurry at the time so they gave us a pass to the front of the line. Everyone was totally cool with that.

Back in the ward, fun could be mighty tough to come by, so we had to make do. I rigged up flying dinosaurs on fishing line wherever we happened to be. Pity the poor doctor or nurse who came into our room. Some of them dropped by just to see what would attack them next. Later a psycho-babe told me that it was a "heavy empowerment trip." I knew that.

At Christmas we made clamshell ornaments. I smuggled a little tree into our tiny cubicle in the clinic. We decorated it all up and made presents for the nurses, checks for one million dollars. That got them good.

Then there was the surreal experience of watching a soap opera about a hospital from inside a hospital. Once a monitor went off in a scene on "Young and the Restless." A nurse came in and wanted to know what the problem was. I told her "Cole's mother flat lined." She cracked up.

That's about as fun as things got in the old Hem-Onc clinic. Working there has to be the toughest job in the world. I really should have made those checks out for two million. They had to deal with us, the hillbilly cancer family, and we were liable to do anything.

Like the time old Clyde cut some red and blue huckleberry brush up the Sol Duc. I kept a cooler full in the truck for a feed anytime. The nurses freaked at first, but it beat a stomach tube and Lulu shared the berries.

Ma read the medical file, which was an unheard of practice at the time, and spotted three mistakes, any one of which could have been fatal.

I'll never forget that last day of chemo. They gave Lulu a kite. We put it together. Picture a little bald kid dragging a kite down a crowded hallway. There's a symbol that took me years to figure out.

Now we're hiking the hallways of our second home, seven stories and several acres of which we know every nook and cranny. We nod to the familiar doctors and nurses. I wonder what they did with those checks.

There is a new crop of bald kids. We wonder what happened to our little buddies. Healed up? Or as Kesey said of his Jed, "Riding point across the river." We don't ask.

The palms sweat until the news comes and it is good. Then we walk out feeling ten feet tall. It's payback time. It is good to be alive.

February 7, 1999

The Olympic Peninsula's Best Fisherman

I may be wrong, but I'm positive. There is nothing more beautiful than snow on a river. We ride the blue water through a white world totally immersed in the rarest thing on earth, silence.

The old bald-headed eagles sit hunchbacked on the frozen limbs. They shake the wet snow flakes from their heads and try to look majestic, but we can tell they are soggy miserable birds by the way they let us float under them without a care.

We are only here so long. Before you know it the drift is over and you're just an old spawner headed downstream tail first. Maybe it would be smarter to be inside by the stove watching the ball game on a day so cold the river smokes, I wouldn't know. I thought we were supposed to catch some fish before the end. I didn't get this way by myself.

My old man kicked when I was a kid. But I got taken out fishing anyway. I was adopted into a clan of the best fishermen on the Olympic Peninsula, the old man of which died last week.

That was a bummer I thought, there was going to be a funeral. The last time we buried an old fishing pal I couldn't recognize who the preacher was talking about. There was an old boy who caught more fish, trapped more fur and killed more game than most people would ever see in a lifetime and all that was said was, "Harry enjoyed nature." That was like saying "Genghis Khan enjoyed camping."

So I wondered about an appropriate service for the greatest steelhead fisherman on the Peninsula. I would have propped him up in his old wood drift boat, stuffed it full of pitch wood, lit the works and launched him off the spillway of the Lake Mills dam at high-water.

And why not? The Elwha River may have been his favorite though he loved them all. The family had a resort on the Elwha just upstream from the Ranger Station back in the old days when fishing was all that really mattered. It was too good to last.

National Park service burned Waumilla Lodge. It didn't fit into the management plan. Neither did my idea of a proper funeral service.

It was okay. The old man knew it all and was not bitter. So he said if we wanted to honor his memory we'd show up on the Allen Bar on the Hoh River. So we did.

This wasn't easy. Highway 101 was a frozen snake of a road. It was a textbook study of treacherous bad weather driving. There was a crowd on Allen Bar anyway. Most were fishermen; some were not.

It was a plunking party for the greatest plunker of them all. Plunking is a stationary art. Instead of stalking the rivers and searching the holes, you sit in one place and wait for the fish to pass and maybe bite.

This sounds easy but it is not. The rivers are huge and the fish only swim in certain channels when conditions are right. These must be discerned through keen observation. You could spend a lifetime trying to figure it all out. Or you could just talk to Paul.

He was a generous soul who shared the secrets of the peninsula rivers from the Elwha to the Quinault. On any given day he could give you the latest report on any of them. Now he is gone and we're left to figure stuff out for ourselves. Still it was funny the way things turned out. Paul fished religiously. The river was his church. The memorial was held on the river. In the middle of the nastiest stretch of rotten winter weather,

Paul made the final call, just like when we were kids. We would fish the Hoh.

Nobody thought we would even be able to fish because of all the storms and high water, but it was like Paul knew. The sun came out. The river dropped into beautiful shape. The old timers Steelhead Club put on a beautiful barbecue.

The beach was lined with plunking rods, but of course the old timers caught the only fish. It was okay. We found ourselves sunbathing on the Hoh River in January. Amazing. If I've said it once, I've said it a hundred times. Thanks for the tip, Paul.

January 22, 1999

"Microsoft spokesman denies rumors Bill Gates is building an estate on Lost Mountain, south of Sequim."

Sequim Gazette, 9/19/99

The Nude Fishing Discount

I remember we were hooking a lot of fish that day. We'd lost every one of them. Usually that's okay. The longer you're on the river the more excuses you have for losing fish. So I never run out. But excuses don't work on the eighty-something crowd. They've heard them all.

I know from fishing my own grandma that you can never trust old people. Any trip might be their last. This granny hadn't caught a steelhead in thirty years. The pressure was on.

Still, we had a nice chat: the Depression, Prohibition, the War, you know, the good old days. Back when no matter how tough times got, there was good fishing and hunting. When the tide was out, the table was set. There was logging too. Why, you could make a dollar a day in the camps. Highway 101 was a single lane of mud through a tunnel of old growth timber.

Now days these kids can't even...Thank the old fisherman, Granny's rod went off before I had to hear how what happened was all my fault. It was a nice steelhead that threw the hook at the top of a three-foot jump.

The crew was mutinous. I was about to reach for the fish club when I saw the kayak. There are few things on earth more worthless than a kayak or a raft on a river. These people are not fishing. They are having fun. There is only so much fun in the universe.

It's tough to fish all day with a skunk on your chest. To finally find a hole full of fish, only to have a kayak come in and execute that perfect Eskimo roll.

Steelhead are nervous critters. They'll scatter and sulk for no reason at all. And here was a yellow kayak with a woman and a dog, just a pup, aboard. She wanted to know if there were any rapids in the river. I assured there were rapids, but no human had ever survived them.

Just then we hooked another fish. Granny rassled it in. Her first steelhead in thirty years and she had to release it because it was wild. It was okay. She blamed me.

The kayak held upstream. She followed through the rapids. The pup got dumped in a bad spot. I offered to take her aboard. The lady assented.

She was built like a brick smokehouse, with dark hair, brown eyes and pearly white teeth. I picked her up and set her in the stern. She sat with regal posture. Do you believe in love at first sight? I'm certain that it happens all the time. With just one look I knew she was it, the ultimate bird dog. She sniffed my hand. She didn't move a paw through the Hell Roaring Rapids.

I was about to say goodbye forever but her owner booked a trip for the next day. Said it looked like fun. There's that word again.

She told me she was an Idaho river guide on "vacation". She liked to fish and bird watch even after I explained how I charge five dollars for every bird I see and two fifty for every birdcall I think I hear. She was loaded.

It turns out river guiding is a lot different in Idaho than it is here in Washington. For one thing, they don't wear any clothes. She said she wanted to "get some sun," and before you know it she was up on the splash deck in the altogether.

You don't see something like that every day. We're fishing the rainforest. A sunny day can be a rare sight. But I'll go fishing anyway. Lucky I hadn't mentioned the nude fishing discount.

We drifted by a couple of bank maggots. I asked if they had any bites yet but they just turned all red and wouldn't say anything. Cat got their tongues I guess.

I pulled into a slot beneath the Rainforest Road. It was a Sunday morning in tourist season. Traffic was brisk. I had just got the gear working when I heard a squeal of skidding tires and the crash of twisting metal. Must have been a wreck.

I was worried the commotion would spook the fish but just then Idaho's rod went off. A trophy bull trout wallowed on the surface like a gut shot carp. A tourist scrambled down the rocks with a camera like he'd never seen an endangered species before.

Idaho wanted to leave just when we started hooking fish. I wanted to hang around. Maybe I could sell another trip. Then she explained. She wasn't really on vacation at all. She was hiding from her crazy rich boyfriend. The guy with the camera could be one of his goons. She was on the run from Possum-Trot, Texas. Wild Bill was on her trail with a helicopter grid search west of the Divide.

So it was another fine mess my big mouth got me into. She was a ton of trouble stuffed in a hundred pound sack. Maybe I shouldn't have offered to book Wild Bill's girlfriend a room at Lake Morning Wood Lodge. But I did. And if he wanted to get tough with the women here in Washington, he'd get a dose of buckshot up his tail rotor for his trouble.

They'd run wild together in the streets of Possum-Trot back in the old days before Wild Bill struck it rich with some kind of electronics scam.

Then things got crazy. "The puppy had an accident on some ratty old rug Wild Bill just paid $800,000 for," Idaho said. I wondered how you could put a price on a puppy's self esteem. "So I left." She described her flight through the western scablands, Wild Bill on her trail with a fleet of helicopters.

I suggested it might be easier to hide from Wild Bill if she wasn't driving a silver Humvee with tinted windows, but she wasn't in the market for any free advice. Besides, I had some logs to yard back home and a Humvee might be just the thing once I cut the fenders off.

Back at the Lodge things went okay at first. Fall was coming. The geese were flying in crazy circles up the valley, trying to find a way south in the fog. Soon it would be bird season and I needed a bird dog in the worst way.

The pup was just starting to respond to hand signals and mark two birds when Idaho panicked. A helicopter swooped in out of nowhere and roared off through the treetops.

"Not to worry," I assured, like the swallows at Capistrano, every fall as regular as clockwork the helicopters return to the upper Dungeness. It's harvest time for the locoweed hereabouts.

Folks can make a fortune growing it. The cops spend a fortune looking for it. It's the biggest cash crop in Clam County. What with a common weed being worth hundreds of dollars an ounce, folks are scouring the hills looking for it. It's like a modern form of the Prohibition, the old timers tell me.

"You get used to it," I said, offering Idaho some earplugs. But she kept insisting that it was no nark. Even at a hundred and twenty miles an hour she could make out the nerdy glasses, the pocket protector with the smiley face, and the bad haircut.

"I gave him that haircut. It was dark. I was drunk. It's just hair. It'll grow back," she tried to explain.

Just then I had a feeling she might be right. A flock of rabid chicken was coming home to roost. Her Wild Bill was our King Billy, the crazy rich plutocrat who bought himself a mountain upriver.

"That's what he does," Idaho explained. "Takes over the water holes and runs off the nesters."

And to think how I rolled out the welcome wagon when King Billy first came into the country. We're practically neighbors. So I hiked over with a plate of sticky buns and a jug of green home brew just to be neighborly.

I figured since he was a newcomer he might not have read this column yet, so I could still sell him a fishing trip. I was really nailing them. Stranger things have happened. That's when I hit the bear fence. I gave it the old hillbilly test to see if it really was electrified. I woke up later that same day. My buns were scorched. The home brew was curdled. So much for being neighborly. Some days it doesn't pay to be nice. The rich are not like us.

So I went back to my side of the canyon and that was that, I thought. No more being nice to anyone. Then the flocks of helicopters started buzzing over at all hours. I should have known they couldn't all be dope spotters. But I could never have guessed the vile nature of the heinous plot that threatens the very fabric of our way of life, if I could just figure out what it was.

I might have done something. Like Paul Revere I could have warned the flatlanders, but no, this Paul Revere was gone fishing, really nailing them.

I came home from a hard day on the river. The Humvee was still hooked to a nice turn of cedar logs. No one was home. The spoor was too easy to read. A chopper had set down in the bean patch. There was a basket of beans and the little dog's collar. And here I was just going to worm her. But it was not to be. The feud was on. Wild Bill stole my woman and my dog. I sure miss that dog.

September 22, 1999

The Olympic Hot Springs

It was going to be one of those days. It's always rough breaking into a job. I wanted to be an award-winning writer. I wrote a column for our local newspaper, a dynamite exposé on compost! Why, I could compost Godzilla in a month if I had enough worms. That's the secret. I even took a bucket of special blend compost into the newsroom just to show them I had the goods. They acted like I had bad breath or something.

My lowlife editor sent me on assignment just to get rid of me. It figures. First he tries to kill me sea kayaking, didn't work. So he sends me to Olympic Hot Springs, where no travel writer has ventured before.

I think he's just jealous because I write good. But here I've been at it for weeks and I still haven't won an award. It's his fault. He butchers my columns until they read like Foghorn Leghorn on acid.

Nobody ever told me you had to pay money to get into Olympic National Park! I'd already blown my expense budget on a box of Chardonnay and a pack of hot dogs.

"Have a nice day," the corrupt federal bureaucrat smirked, as she skinned me enough to pay for a month of cable TV just to get into this crummy park for a year. And you can't even change the channels when it starts raining. It was easy for her to have a nice day; she was obviously senile from being locked in that booth answering tourist questions. I could tell. She was just a little too nice.

As I drove further into the hills my talk radio station was drowned out in the static. It was okay. They were talking about traffic jams and I was in one, crawling behind a conga line of gawking idiots that wouldn't pull over no matter how much I flashed my lights and honked my horn.

Just as I was about to pass, I came to the end of the road. That was the good news. The bad news was the hot springs were another 2.4 miles and the road was barricaded. I might have run it, but there was a gang of surly looking young people replete with extensive tattoos and pierced body parts standing in the way adjusting some backpacks they'd probably just stolen somewhere. What sort of people do they let into a national park these days anyway? And here I'd forgotten my pepper spray. It was okay. I was ready for them when they started whining about what a nice day it was. What is with some people? The sun can't come out for 15 minutes without them making a big deal about it.

Seething, I filled out Big Brother's backcountry permit like it's any of their business where I go to get lost. The hot springs pools are contaminated with a variety of organisms with big names, the kiosk informs. Just sort of figures.

This is the end of civilization as we know it. I walked up the road. Moss and kinnickinnick creep along the edge of the pavement. Creeks have washed out jagged sections; the trees are taking over.

The road has only been closed a few years. Shows you what nature thinks of us.

Every little while I met some hikers coming back from the hot springs. I thought it was strange the way they all smiled and tried to talk. At first I thought my fly was open. Then I resigned myself to the fact there are a lot of weirdos in the woods who think passing on a trail is an emotional commitment.

I made it to the campground and collapsed on a mossy rock. I did not remain collapsed very long. The whine of bloodthirsty insects was deafening. I squashed a bloody smear all over my face and neck. I got up to set up the tent but found I'd forgotten the poles. I tried to build a fire but dry wood was as tough to find in that campground as $20 dollar bills on the streets of Seattle.

All I could find were the exposed remains of many a morning's ritual. Archaeologists have somehow determined Cro-Magnon man stopped soiling his cave approximately 40,000 years ago. This hygienic revolution had yet to reach the Olympic Hot Springs campground.

I thought I should check out the polluted hot springs. I was a professional on assignment. I crossed Boulder Creek on a rustic log bridge. I smelled something, sulfur, and then saw steam rising from a creek running down the trail. I soaked my aqua socks. I felt the heat from the water. It was good. I came to a pool with a spring gushing out of a little cave. The water seemed perfectly clear. I settled in under the little waterfall. The temperature of the water was adjusted to perfection. I wondered how this wonder of nature could exist in an imperfect world. Forget that plate tectonics stuff. It's just a theory no two eggheads can agree on. I'll go with the Indian version:

Long ago there were two dragons. One dragon lived on the Elwha, the other on the Sol Duc River. One day they met on a ridge and engaged in a mighty battle that lasted for years. Neither dragon could slay the other. They knocked down all the trees in the high country and ripped big chunks of hide off themselves until each retreated underground to his own valley where the tears of their pain gave us the hot springs.

Into each life a little rain must fall. Then it began to come down in glorious sheets. It was there I discovered one of the true pleasures of life, to sit in the springs with the rain beating on my face.

I turned to soak my feet in the waterfall and floated on my back looking at the world upside down. The pool was surrounded by a hanging garden of ferns and wildflowers. Spiders' webs were covered with tiny droplets of condensation until they appeared jeweled. The place looked like it had been decorated for Christmas. My gaze faded to a timbered ridge of the most beautiful fir timber God ever grew. A pair of goshawks circled above. My spirit soared with them.

I sensed movement. Hominids entered the pool appropriately disrobed according to local custom. They spoke but I could not understand. The waterfall was too loud and my ears were under water. It sounded like Italian. I tried my limited facility in that language, "Pass the wine, pretty girl."

They appeared confused if not alarmed. They were Israelis, so all I could say was, "The coffee is good in America." It was okay. They understood in their own way. There was much nodding and smiling in the twilight. When they lit the candles on the rocks, I felt it was time for me to leave.

No one had set up my tent in my absence. I noticed a fire nearby. I wandered over just to be neighborly. By some coincidence, it was the kids with the tattoos and the nose rings. Suddenly I was glad I hadn't run them over. They were having spaghetti and garlic toast and they offered me a plate, delicious. I stabbed my box of wine and poured it round. They said they had an extra piece of plastic and some string if I needed a place to sleep. I had to borrow a flashlight too.

I slept that night the sleep of the dead. The sound of the creek washed the noise from my head. I awoke, and thought I should write a poem. The hot springs beckoned. I returned to my place, feet in the waterfall, head in the clouds. I remained until a pang of remorse rent my conscience. My dear editor had sent me to these springs so I might share this wonder of nature with the gentle readers of our friendly little paper.

I pried myself from the water at noon. I packed up and said goodbye to my tattooed friends. I hit the trail to town. Others were walking up the trail. I tried to speak to them but they just looked at me funny. I checked my fly again. Then I understood. These suffering people had not been to the hot springs yet.

I drove slowly back to town, stopping once for a baby bunny to cross the road. You really can't be too careful. I wanted to stop at the fee station and tell the nice lady how much she reminded me of my own grandmother, but she wasn't there. That was okay; I would send her a thank you card later. I went to town and had lunch with my editor. It was a good day to get a tattoo.

May 13, 1998

"National Guard Sent to Neah Bay for Makah Days"

Seattle Times, 8/11/98

Saturday Night in the Fifth Reich

"The pen is mightier than the sword," my jack-booted editor chuckled from the hot tub on his redwood deck where he was brewing another screed on the global destruction of the rainforest. "I want you to cover the whaling thing."

So I was off to Neah Bay for Makah Days. I wondered if the pen was mightier than the Ml tank, or the Bradley fighting vehicle, or the M-60 machine gun, or the Apache Blackhawk helicopter, or the combined forces of the National Guard, FBI, BIA, Coast Guard, state patrol, and sheriff's department we saw convoyed west through Sequim this week. I hate to admit it but the little creep was right, and the fact that you're reading this and I'm not in jail yet proves it.

"Don't get scalped," he giggled the stereotypical ignorance of a self-obsessed baby boomer know-it-all who thought the history of the world began with the invention of television.

Shows what he knows. The Makah never scalped anyone. According to old Judge Swan who did ethnography of the Makah for the Smithsonian back in the 1860's, they were headhunters.

Once I asked a carver where he got his ax. It was a rusty little thing that didn't look like something you'd find at the hardware store. He claimed one of the Grandpas got the ax from a Spaniard and lopped the guy's head off.

I asked him why the Indians didn't kill all the white boys that washed ashore. He gave me the Indian answer which was no answer at all, an answer in itself. Maybe they were just too nice. I doubt it. Back in the age of exploration folks sailed here from all over to get massacred.

The real reason the Indians didn't kill all the white men back when they had the chance was trade. Whites wanted furs. The Indians wanted metal. This set off a genocidal orgy we like to call the fur trade. Both sides got more than they bargained for. For the Indians it was a cultural grab bag that included smallpox, house trailers, and the Bureau of Indian Affairs.

Smallpox hit the Makah in 1852. They died so fast their bodies were left on the beach to wash away in the tide. In 1855 the Makah traded again, approximately 700,000 acres of the world's most valuable timberlands, for a reservation.

The treaty of 1855 let the Makah kill whales. By 1920 this was like telling them they could kill buffalo. There weren't any whales left.

This is the nineteen nineties. The whales are back. They are friendly, curious, and trusting. Which is how I feel about them any time I've been out rowing or paddling close to whales in a small boat. Whale watching is an industry. The Makah were set on whaling and everybody else was against it. It was a media frenzy. I thought it would be fun.

I headed west on State Route 112, which according to local legend was laid out by a logger tracking a bull elk in the rut. The path of true love can twist and turn a near circle, which is just what 112 does. But State Route 112 is more than just a scenic drive. It's a textbook of environmental and human degradation. I like it.

It starts by crossing the Elwha River on a high canyon bridge. According to Elwha S'Klallam legend the first man was created just upstream from dirt in pools in the rocks. Until 1912 you could foretell the future by throwing deer hair in the pools. Then Tom Aldwell built his dam and flooded the place. We've had to rely on the psychic hotline ever since. There was no way for the salmon to get over the dam. The salmon died out for some reason. We're still studying the problem.

I crossed Deep Creek, where I caught my first steelhead. It's where my 80-something plunking pal caught his first steelhead too. My kids won't be fishing Deep Creek. They logged her too hard in the 8O's. Deep Creek is permanently closed to fishing.

I drove through miles of cut-over land and eroded hills, over barren silt-choked streams, through scenic fishing villages where fishing has been outlawed. With the timber gone and the fish extinct a new prison has been built, offering the only hope of steady employment to those hardy souls still hanging on.

It's all part of an environmental catastrophe no one seems to notice. Then the Makah wanted their treaty right to kill whales. Suddenly everyone's an environmentalist. The squads of motorcycle policemen told me I was getting closer. The roar of black helicopters, the jam of Humvees and troop carriers had everyone in such a festive mood I didn't mind paying $5 dollars to park. I've paid to park in worse places.

I used to fish Neah Bay. I thought it would be a homecoming. Forget that. It was a Saturday night in the Fifth Reich. We were all under surveillance. Every other Anglo was a reporter or a cop. There was no riot, no story, no bars. The cops were very bored. The reporters were very sad. I saw two reporters interviewing each other.

It was okay; the kids were having fun. This was their big carnival and no one was going to ruin it.

Following my reporter's instincts, I made a beeline for the big purple Barney jump-o-rama but I couldn't get in, stupid height restrictions. I waited around for an eco-terrorist to run up and harpoon Barney. It would have been a symbolic act. Worth a picture. The public perception of the Makah harpooning a whale.

The vendors had a fine display of whaling T-shirts you won't find anywhere else. The kids performed traditional dances on the main stage. A gaggle of media scum, yummy info-babes, and big-haired guys had a tribal official cornered against a pile of driftwood. I moved closer to listen, but heck, I've spent too long on a chainsaw, couldn't hear a thing except the word, "money," repeatedly.

Finally I could relate. I thought if it was just a question of money, the Makah should do what we fishing guides did, catch and release. Just touch the whale, count coup, and charge people more money for not killing it. It works for us guides. For one thing, you don't end up cleaning a bunch of fish or whales after a hard days whaling or fishing.

That Saturday in August in the '90s at the peak of tourist season I walked to the tip of Cape Flattery, the most Northwest point in America. I was alone. I drove south towards Shi Shi beach, past miles of the most beautiful deserted coastline in North America.

I wondered why the Makah tribe wouldn't market this scenic splendor to the tourist industry. I would. Why couldn't the Indians be like us? We were only trying to help. We gave them the fur trade. Then the fur was wiped out. We told them to farm, though no one ever made it farming here. Then we told them to log, until the logs were gone. Now we tell them to be tour guides, an even bleaker road to nowhere. So we built them a nice new prison at their doorstep. It's one of the biggest growth industries in the country. You'd think the Makah would appreciate that, but no, they'd rather be whaling. What's the matter with those people anyway? Why can't they be like us?

August 12, 1998

The Bear Problem

It was another tough week in the news. Tourists harassed abandoned seal pups on Dungeness spit. A plague of chipmunks invaded a local home, and what? Bears closed Olympic National Park. Even the Seattle news stations dropped coverage of the latest gang related drive by sex crimes to cover this one. Okay, so maybe the bears didn't entirely close the park, maybe it was just eight out of six hundred miles of trails. There was film at eleven. Not of the offending bears, but then, they all look the same. This was a crisis in the making and it's never too late to panic.

My power mad editor called in tears. He knew I was the one to cover this story and it killed him to ask me. Just because I've badmouthed him for years he's suddenly hypersensitive about it.

"Try to put a millennium spin on it," he whined through the suds of his latte. "How about 'gang related'?" I suggested just before he hung up on me. Finally here was a story I could sink my teeth into. I've been camping with bears for years.

Once upon a time, before I was an award-winning writer,[1] I had a real job thinning trees on top of Bonidu Mountain.

The bears were thinning trees too, only not with chainsaws. They stripped off the bark, sometimes the full length of the tree. Then one

1 Pat Neal has never won an award for anything. Editor

day there was what was called a "damage control" hunt. They wasted the bears for doing the same thing I was getting paid for, thinning trees. Talk about downsizing! Lucky for me I was in the right union.

The Indians said bears were the mothers of all animals. They killed more salmon than they could eat, sometimes just stripping the eggs out of the females and leaving the rest for the lesser critters to feed on.

So now with the salmon gone, who could blame the bears for switching to camper's grub. Not that it's a fair trade. Ever had a fresh spring salmon? Compare that to your freeze-dried sludge. Sure the bears will eat it, hermetically sealed foil pouch and all. But that doesn't mean they'll like it.

Then it hit me. It is a gang related millennial thing! The bears are trying to tell us, "Let fish get back up the Elwha or we will shut down your precious park."

It all made sense as I prepared a light supper over the campfire in the twilight. Just a few scampi with asparagus and Hollandaise, a truffle risotto, and chocolate mousse for dessert. I could smell the musky, sweet smell of a bear nearby but I wasn't worried.

To avoid trouble with bears I try to act like one. You know, the snoring, scratching, and stuff. Bears are solitary animals except for the breeding season and as you can probably guess it's been a long time since I had a date.

I tried to leave a clean camp before I went to sleep but then I tried to remodel a bathroom once. Some things you just can't fake. I awoke in the darkness. She was licking the chocolate mousse from my face.

May 27, 1998

The Mountain Goat Problem

I've never caught a fish while mowing the lawn. My editor loves to mow his lawn, but then he doesn't fish. I found a better way to deal with the problem. Park the truck on it. It's a toxic waste dump on wheels. The lawn turns to lovely shades of orange and black you don't have to mow. Housework is another complete waste of time. Sure I don't mind hosing off the dishes or shoveling the floor every week or so, but I don't do windows unless one gets broken.

So when it came time for some heavy cleaning I knew just what to do. I loaded the brat pack into the wreck and headed for the high country. Dehorn Lake was the call. We were looking for mountain goats to see if any had survived the purge.

Before you go looking for Dehorn Lake on a map, don't bother. Few of the names in this column have been approved by the Geographic Board of Names. It's an old guide trick. I make up names for everywhere. That way when clients brag about where they shot this or caught that no one will have any idea where they are talking about. They can't nark off the water.

"I caught a twenty pound buck steelhead in the Spandex hole," I heard one unfortunate foreigner declare in a crowded bar in Forks. The patrons left him alone after that.

It was ten years ago today we soaked our bleeding feet in the icy waters of Dehorn Lake. There was a herd of 19 mountain goats on the cliff tops above. Watching goats on a mountain was as much fun as climbing it. They lived on ground so steep a human would need a rope, a death wish, or both to venture up with them.

We were having a wine and cheese tasting party. I was cutting the cheese when the goats stampeded and scattered across the mountain. Then we heard a cougar scream. The cat had blown its stalk and was really mad. Goat meat is tougher than a boot, but meat all the same. The goats stood as still as statues. They knew nothing could catch them. Sure.

Ten years after, we hit the trail to Dehorn Lake. The trail forked and the kids took a wrong turn. Or maybe I did. The trail fizzled into a shin-tangle of huckleberry brush that hung down the mountain like matted dog hair. We were lost.

Whoever said, "Don't panic!" when you're lost, has never been there. I never miss an opportunity to panic. We immediately sat down and ate all our chocolate. I knew there was a lake up there somewhere, so we kept climbing until we were looking down on it.

We climbed down to the lake for a swim. I was almost relaxing when a car alarm went off and would not stop. We were miles from a road. I was confused until Lulu explained it was a whistling marmot. I knew that. Marmots whistle when they are alarmed. They are alarmed a lot. Everything wants to eat them.

We'd hiked through acres of marmot condos, cozy little burrows. Some had water frontage, each had a sun porch and a terrific view. We'd noticed little stacks of mountain herbs drying in the sun. The marmots were making hay. Marmots are one of the few animals around that will eat a Devil's Club. It's like a rainforest cactus. You'd have to be tough and hungry to eat one, which might describe a marmot's winter in a burrow twenty feet beneath the snow.

Now it was summer. No time to worry about snow. I had a bigger problem. It took the kids about a minute to figure out there were no mountain goats at Dehorn Lake. I had lied. They were madder than when I got us lost.

I had to explain how a bunch of government do-gooders evicted the mountain goats from the Olympic Mountains because they were an introduced species. Like it was the goats' fault. The mountain goats had been, "relocated", as Goebbels would say. Must have been easy. They would do anything for salt. Dump some anywhere and you had a mountain goat party. The party was over.

There was a pang of lonesome silence on the shores of Dehorn Lake. Life is change. I knew when the goats were gone the fishing guides were next.

We were lured to a meeting at Kalaloch. There'd been a rumor of free drinks. Another introduced species bites the dust. There'd been no mention of "relocation." Just a gob of new rules no one but a lawyer could figure. If you cannot afford an attorney, you probably can't afford to go fishing. It was okay. The worse fishing gets, the more you need a guide.

July 22, 1998

The Elk Problem

And so another elk season passes astern. People used to come here from all over to hunt elk. To get ready for this hunt of a lifetime I dropped by the local supermarket and picked up a few glossy magazines that showed how to bag a trophy elk.

I read many of these articles about elk hunting, trying to copy one for this column. First you ride your faithful horse at the head of your pack string to a tent camp in a remote copper washed canyon in the Rocky Mountains. The faithful guide lights the stove in your tent before you have to get up and enjoy the heart stopper breakfast prepared by Hans the Chef in the big dining tent.

On your stroll to the banya you spot a trophy elk but it's not as big as the one your buddy shot, so you wait for a bigger one. And sure enough you get a great big elk. Nice shot! The faithful guide wrestles the carcass back to camp in time for cocktails.

I know these stories because I've heard them from my clients, like they expect me to do all that stuff. And I will, the day after I grow an extra pair of arms.

Elk hunters have so many unrealistic expectations. Like, they want to shoot an elk? They are not happy when I explain that for the second time in this century the Olympic Elk have been hunted out.

Sure there's a herd of refugee elk applying for environmental asylum in the city of Sequim. Not everyone is thrilled about it. It's just another case of blaming the victims.

The elk were up in those hills for years and nobody in Sequim knew it. But this is not a perfect world. We turned elk country USA into a combination county dump and shooting gallery. The elk found it tough to adapt to a diet of discarded wallboard, car bodies, shot up barrels of waste oil, and sacks of medical wastes. They were allergic to high velocity lead. The elk moved to town for some peace and quiet.

No one has ever written much about hunting elk on the Olympic Peninsula. Could be many reasons for that. It's tough to ride a horse through a clear cut. Many of the best elk camps are trailer trash in gravel pits and big city editors don't think that's picturesque. And it generally rains a foot or so during elk season here, making it one of the most truly miserable experiences you can enjoy on this planet.

But now that the elk are gone I can tell you how to get one. The elk have always loved loggers. Think about if you were an elk. Things are really boring out in those woods. Then some guys show up and knock down a bunch of fresh browse and scrape up some new dirt. There's noise, excitement, and the smell of burning diesel. The elk would stand around for days watching loggers. What else did they really have to do?

I used to dress all my elk hunters up as loggers, high-water tin pants, hickory shirts, suspenders, and hard hats. We'd head out early and find a fresh logging show. I'd build a big fire, a rigging slinger's fire. We'd stand around and chew and cuss and spit, or toss our hard hats at an ax handle for money, or throw the ax at a log until we broke the handle.

Every once in a while I'd start the power saw. I was guiding after all, and cut some more wood for the fire. The sound of a really big chainsaw was like a love call to an elk. And before you knew it, there were the elk coming to the decoys.

You wanted to be careful about shooting at elk around loggers. If you accidentally shot a logger and he found out about it, he was going to be mad. And I wouldn't know you.

Luckily most of my elk hunting clients were bow hunters. It was a curious fact that many of these bow hunters had never seen a real elk in the wild. They shared the belief that it was too easy to shoot an elk with a gun. They were going to give the elk a chance and get one with a bow and arrow.

I think archery hunting is just too easy. If you really want to give the elk a chance, hunt them with spears. That's the way the Indians did it. They were in shape.

You just run a big herd straight up the side of a mountain. The elk in the front get bogged down in the brush and the ones in the rear can't get out of the way on a narrow trail. Go ahead, poke that thousand pound bull elk. That's giving them a chance. It might be fun to go spear hunting again, if there were any elk.

November 18, 1998

Checkers the Cat

It was another tough week in the news. Although, if you think you have it bad, try writing a nature column. The first thing you'll need is a little peace and quiet. Good luck. Between the howling packs of coyotes, and the dueling owls, there's little opportunity for sleep. The ducks start quacking at daylight in the swamp. The chores go on forever.

Just when you'd think things couldn't get any weirder, Checkers shows up with a friend. Checkers is my pet cougar. I know, I know, everyone says I should shoot the cougar. Then what? Do you have any idea how hard it is to find a good pet these days? And Checkers is better than most pets, your average house cat, for example.

I know this is hard to believe, but some people keep cats in their houses. On a recent trip to the evil city in the east, Seattle, I saw first hand the results of harboring these vermin in one's home.

The first thing I noticed was the smell, the overpowering aroma of cat. I've smelled this in the woods before. Checkers will rake up a pile of moss and dung. That's how cougars mark their ground. Ignore this sign and you could get eaten, if you're a cougar anyway. Cannibals.

It's one thing to find a mess like this in the woods where it's a wonder of nature thing, but I had no idea people kept this stuff in their house.

And not just anyone's house. These are the beautiful, intelligent city people with big brains and fat wallets. Nothing at all like the hard core of shiftless hillbillies you'll find swilling at the Lodge on a Saturday night coon hunt. No, these are my fancy friends. They have electric this, automatic that, and cat dung in the kitchen.

"It doesn't smell bad does it?" my fancy friend asked. Heck no, my eyes always water like this. I sat on a couch that probably cost more than my drift boat while the cat raked its claws on the other end of it. I hit the cat with an ice cube. It ran into the kitchen and jumped on the table, scattering a plate of hors d'oeuvres. Something must have scared it.

Checkers has never once been that much trouble. But these same people will come up to my house and insist I shoot her the minute they see a fresh track. It's an outrage!

Checkers has a litter box, sure. But it's out in the brush somewhere. I've never had to worry about it. Checkers works hard at keeping her claws sharp too. That doesn't mean she has to rip up my fine heirloom furniture doing it. Not while there's plenty of cedar trees. She rakes the bark right off them, twenty feet in the air.

And I don't have to worry about what kind of food she might like. She's got that figured. Most any walk in the woods reveals the freshly disemboweled remains of Bambi or his mom.

The first thing you notice is the hair. Checkers licks the hair off a deer hide before she ties into the big feed, usually starting with the paunch.

You should try to rip the hair out of a fresh deer hide sometime. It's a tough job. Look at the pile of hair around a cougar kill and you'll see the nature of the beast.

Checkers likes to hide her kills under a pile of brush. It's a good idea to hide your meat from the ravens. It takes ravens about two minutes to find something after it dies. Then they just have to tell everyone everything. Sometimes, the ravens will tell you where the cougar is.

It's fun to hang around a cougar kill. It's a great way to get some cool antlers. Which is about the only thing that doesn't get eaten. Don't worry; cougars don't usually guard their kills. It's enough to know that Checkers is nearby. It's a warm feeling, like you're never really alone.

In the good old days a pack of wolves could come along and run a cougar off its kill. Now the do-gooders want to bring the wolves back. The cougars will not be happy. The locals will not be happy.

But just remember, the folks who say they're going to bring the wolves back are the same pack of slackers that said they were going to bring the hundred-pound king salmon back to the Elwha river. They studied that problem until it went away.

For now it is enough to know that it is spring. The old cougars' thoughts turn to love. I've heard them yowling. Their tracks follow mine.

February 14, 2000

What if you take your kids fishing too much?

A hundred years from nowhere.

A cougar kill. It's a great way to get some cool antlers.

He's just an old spawner.

I think I've always had a fishing problem.

The end of the last frontier

Enchanted Valley is.

The Elwha River might have been his favorite.

It's a wild thing that will kill you.

Wally the Weasel

It's springtime at Lake Morning Wood Lodge. I can tell because my pet weasel is in heat. Wally the weasel is his name. He's about the only pet I have left. The cougars ate the rest of them.

Wally showed up one evening in the middle of the worst mouse infestation in years. I couldn't trap them fast enough. I tried shooting the mice. But you know how kids complain about every little thing these days, the noise, the smoke, the lead dust. Toughen up.

So when Wally showed up riding the lazy Susan, I knew I'd found a friend. I wasn't sure how he got into the house, or how he was getting out. And I really didn't care.

This old shack has more holes in it than one of my fish stories. You can fly a kite indoors on a breezy night. The door is always open. You never know what's going to hop, crawl, or walk in demanding something, a meal, directions, the fishing report.

A weasel is about the neatest looking animal in the whole wide world. There's the sleek brown fur, and a white throat patch that makes him look like he's wearing a tuxedo. He's got a cute little pointy nose, with whiskers, and these little black beady eyes that are just cuter than cute.

And now he has a girlfriend. And what's wrong with that? Wally is a mouse- killing machine. He's killed more than a family of weasels could possibly eat. He leaves the mice lying around for me to sweep up for a proper cremation in the cook stove. No problem. I can do my part.

I've had worse pets. Turkeys, horses, and civet cats make bigger messes. The fact is, I wasn't sure I was ready for two weasels. That's a lot of responsibility for someone like me, having a fishing problem and all.

And besides, weasels are among your more nervous pets. At my house, if you think you see something out of the corner of your eye, you probably do. Get a couple of weasels bouncing off the walls and you can go cross eyed trying to keep track of them.

Still the weasels are very friendly. Or they are rabid. With weasels it can be hard to tell the difference. That's the great thing about wild pets; you don't have to take them to the vet.

A weasel is about the most vicious animal there is. If weasels were the size of bird dogs no one would go out in the woods. They are always hungry. Your average weasel kills fifteen times its weight in raw meat every twenty-four hours. Okay, so I really did make that up. But the truth is they cleaned the vermin out of the house in short order.

That's when Wally moved outdoors to conquer new worlds. He tried to take on the bird feeder. Poor little guy. Couldn't have it was the toughest bird feeder in the county.

One day I tacked the bird feeder to the kitchen window and like an idiot, nailed up some tallow to help our feathered friends make it through a tough winter. The cougar ripped the tallow and the bird-feeder right off the windowsill the very first night. I decided to move the bird feeder before he came in through the kitchen window. He might get hurt. Every animal in nature, from the lowly earthworm to the humble fishing guide, has its own territory. Wally should have known he was moving in on someone else's turf when he headed for the that bird-feeder.

It really shouldn't be called a birdfeeder at all. It's more of a toxic waste dump for experimental junk food. The truth is, I've been feeding the birds the soggy remains of the junk food people gorge on trying to keep warm in the front of the drift boat.

Pork rinds, donuts, microwave burritos, it's one thing to buy and eat this stuff yourself, but feeding it to the birds might be cruelty to animals. I'm not saying these birds are fat, but they have to be really careful about breaking stuff when they land.

What began as a misguided attempt to share nature's bounty with my feathered friends quickly degenerated into a shocking example of the shameful effects of a human diet on animal behavior. Add a rabid weasel in heat to the mix and you've got a story that may not be fit to print.

April 15, 1999

Backpacking

I thought I was going to die. My heart was pounding out of my throat. Breathing was shallow and difficult. I was covered in sweat. It felt like a heavy weight was squeezing the life out of me. Heart attack? Nope, I was backpacking.

The straps of the burden cut into my shoulders like piano wire. I could feel a crop of blisters growing with each excruciating step. It goes without saying we'd gotten a late start. It was the hottest day of summer. It must have been ninety, which is thirty degrees past where I melt. I was bred for cold weather. When that bright thing comes out I anchor up in the shade. This was another mess my big mouth had gotten me into.

My editor and I had been gorging on waffle fries and gravy at the pancake house. He wanted to know what was so enchanted about Enchanted Valley. It has been called the "valley of ten thousand waterfalls." They cascade four thousand feet down an alpine escarpment into a pristine valley with a restored pioneer lodge in the middle of it. When the sun heats the rocks, the rocks heat the water, creating one of the world's largest sun showers. Soak in the showers; take in the view. In a land where half of the natural features are named after bears or whiskey, somebody finally got the name right. Enchanted Valley is.

I collapsed in a heap at the trail register. The gravity of the wilderness experience hit me. I'd left my espresso maker in the truck. What with the trailhead parking lot being full we'd been forced to park on the far side. It was a hike in itself back to the pickup. I whined at the kid until she agreed to go back and get my survival gear.

"Coffee is a drug." Lulu sniffed. It's amazing how mouthy kids get when they know you can't catch them. "Daddy needs coffee," I explained, "or he'll be cranky in the morning."

"You're already the meanest dad in the world," she said, taking my keys. Now I was worried. Like an idiot I taught her to drive when she was eight. I figured the odds were even she'd start that truck and drive her teenage attitude to town. She didn't though. Maybe it was because I had a pack full of chocolate and her fishing pole. Maybe I did something right.

I was badly dehydrated. I found my water bottle full of peach schnapps, warm. By now it was time for lunch. I had been in a big hurry when I packed. I don't know why I thought takeout Chinese would go good on a backpacking trip. The Kung pao chicken was gamey. The oyster beef was too oystery. The fortune cookies were smashed and crumpled. Mine read, "Yours is an enchanted journey."

Then I remembered I was a professional on assignment. I struggled to my feet and read up on the rules of the wilderness. There are many. The first being you will pay to camp up that Park trail. If you think fishing and camping with your kids is some kind of cheap alternative to a real vacation, think again. The true costs have yet to be determined. Unless my tight fisted mean spirited editor starts coughing up the expense account I deserve, I'm going to do something less expensive than camping and fishing next summer. Maybe we'll hop the Concorde and do the Paris Disneyland.

So we hit the trail. It wound through a forest of trees the size of a Winnebago, so tall you could not see the tops. We crossed the Quinault River on a log bridge high over a magnificent canyon. The Quinault has more elk than anywhere. We could see their deeply rutted trails coming off the side hill.

Towards evening we dropped down to the river and found a cushy sand bar with a good fishing hole. A big herd of elk had just been there. We could tell because every inch of beach was covered with droppings, hair, or both. The elk flies, they're like horse flies on steroids, attacked in waves. We fought them off with willow branches until Lulu figured out how to set up the bug proof tent. Just in time. We were running low on blood.

It was okay. Lulu caught some fat trout. We cooked them on a flat rock over the coals. I sat by the fire picking elk hair out of my cocoa, whining how we might as well be camped in a barnyard, when Lulu put her finger on it.

"Dad, there are no cigarette butts." She was right. There were no beer bottle caps, or plastic or styrofoam, none of the signs of man we find littering the wild lands outside the Park. Suddenly paying to get in wasn't such a bad idea. I slept out under the stars. I awoke to a refreshing drizzle.

July 22, 1998

Bird Watching

It was going to be one of those days. Here I was leaving on a hunting trip and my vegetarian editor calls up and wants an article on bird watching. No problem. Right then I was watching a Canada goose circle a bed of wild rice pilaf. I didn't mention most of my bird watching has been down the barrel of a twelve gauge. Old Audubon himself was a heck of a wing shot. He once shot twenty-five pelicans so he could draw a picture of a dead one. The best bird watchers I know are a pair of wetland ecologists. They're both duck hunters. One day I stupidly agreed to go for a little walk in the swamp with them.

Competition Bird Watching is a tough sport. You don't need a helmet but you'd better have a thick hide and a good set of eyes. The object is to spot and identify whatever it is first. I found accuracy counts more than speed when I called a dowitcher a jacksnipe. I was laughed out of that swamp and never went back.

So I was headed for one of the hottest bird watching spots on the Olympic Peninsula, Avalanche Canyon in the Buckhorn Wilderness. It was just a coincidence I had some guy along who wanted to shoot a buck. It was another low-impact wilderness adventure where we take only pictures, leave only a gut pile.

A walk through Avalanche Canyon is a walk through the past. The glacier just melted the geologic day before yesterday. The mountains are too steep to hold themselves up. They tumble down the valley plowing it into fearsome tangles of broken rock and smashed trees. We followed deer trails, but then deer aren't very tall so, when in Rome. We abandoned the bi-pedal thing and bulled forward on all fours.

At one point I could have sworn I smelled rancid kitty litter. We looked around and found a cougar den in a cave. It was a hot day. I'll spare you the details. But momma cougar was a good housekeeper. Baby cougars do not come housebroken. Ma had scraped an entire summer's toilet training into a neat pile and covered it with moss and sticks. That's how adult cougars mark their ground. Ma was teaching them.

Ma was a good provider. You could tell by the scraps of hair and bone around the place. None of it looked hominid, but then I didn't look too close. A nervous looking marmot came out of its den not fifty feet away and whistled an alarm. Now I've seen everything. The marmots bed down with the cougars. That marmot had to be lucky or smart or both. Marmots must have no sense of smell or they would never live next to a cougar's den.

We limped above timberline and camped. I fluffed up my sleeping bag in a dried- out bear wallow. A lone raven flew over at dusk. So far it was one dud of a bird watching trip. It was okay. My hunter was a cop. They're good shots. So we could at least get some meat. He's the sniper on the S.W.A.T. team. He raids crack houses on Saturday nights for something to do. Says it's groovy.

He got a nice buck first thing in the morning. I rolled the insides out and got the meat back to camp. Then we started bird watching. The ravens came in first, an even dozen. It was a party. They filled up, watered at a spring, and flew down to camp to be sociable.

Ravens are great mimics. Things got weird. They started doing impressions of sand hill cranes, Canada geese, stellar jays, and a curious "meow." It was a concert in a thousand acre landscape painting that changed colors with the play of light. I sank into my bear wallow and watched the stars appear.

The ravens were back first thing in the morning. Things went okay for a while. Then a big black shadow came over the ridge. I put the nine power on it, a golden eagle. This is a rare sight in the Olympics. It made a pass and scattered the ravens, then settled in for a feed. The ravens didn't scatter for long. They dive-bombed the eagle until it took off and soared down the valley.

The eagle was hunting marmots. This is tough to do with a flock of ravens on your tail. So the eagle sailed to the next valley, but I had my sighting, confirmed. I'd show those know it all eco-people. Nothing like a good bird watching grudge. Whoever dies with the most species sighted wins. I almost didn't need to see the Clark's nutcracker or the loggerhead shrike, but I did. We're bird watching. We showed no mercy.

September 23, 1998

Canoeing the Dungeness

"Do something about canoeing," my editor screamed as he slammed the phone in my ear. What can you do about canoeing? People have been drowning in canoes for years and they're still perfectly legal.

We're running the upper Dungeness River. It's a wild thing that will kill you. It's the fastest dropping river in the continental U.S. We guides love to pander these meaningless phrases and why not? Some of them just might be true. Like the one about the canoe being the most dangerous or the safest craft afloat, you decide.

Canoeing the Dungeness is not something you want to do alone. But it's hard to find someone stupid enough to do it with you. Luckily a lost Alaskan dropped by the lodge. He'd never seen the Upper Dungeness so he agreed to canoe it. The lost Alaskan's timing was perfect. By the end of July the spring melt had slackened off but there was still enough water to cover the rocks.

We pulled into the Two Forks Campground where the Dungeness and Greywolf rivers plow together. The sun was on the water, which is a good sign on a river of melted snow where you know you're going to get wet.

We launched the 17 foot aluminum canoe and knelt on the deck getting the sea legs, discussing strategy, praying. We agreed to backpedal, to slow ourselves down enough to get some steerage. This plan worked real well for about fifty feet until we slammed into a granite boulder the size of a Volkswagen.

Whitewater canoeing is a team sport. Communication is the key. I didn't know Alaskans had different words for right and left. Since our method of steerage required us to paddle left to go right and vice versa, I never knew what the response to my gently shouted suggestions to right or left would be.

In this state of confusion we shot into a convoluted boulder patch I could not see the end of.

Right or left, it didn't matter as we flipped over one rock and sank upstream of another, pinned to the bottom by the force of the current. We made it to shore and salvaged some gear that hadn't sunk or floated away.

The lost Alaskan was a big lunk, strong as an ox and twice as smart. By working together up to our necks in water we could not move that canoe an inch. It was a long hike out of that canyon and a longer ride back to the lodge. The once proud craft had been recycled into 80 lbs. of aluminum litter on the bottom of a wild and scenic river.

We were back the next morning with a hand winch and enough rope for a high lead set up worthy of the glory days of logging. When we finally inched the canoe off the bottom it was not a pretty sight. She was caved in like one of the lost Alaskan's empty beer cans. We found some round rocks and beat her back into the rough shape she'd started with.

We toasted our good fortune and hit the river again. This time things were different. The impact of striking that rock had indelibly etched the concept of left and right into our collective consciousness.

The rest of the trip is a blur of white water in green canyons. I remember waves sweeping us chest high. Being on end in the stern, watching the bow man plunge into froth, I soon followed. At one point the river hooked a 90-degree turn to starboard. We went straight. I lost the lost Alaskan for a while. He was swept down river screaming, paddle held aloft, a born waterman.

We shot around a blind corner. A load of timber blocked the river. We jumped from the canoe just before impact and except for some bleeding, made a perfect "10" landing. Didn't matter, we were in for the dirtiest French word I know, "Portage".

We could have used the ax to chop through the slash in the logjam if we hadn't already lost it in the river. We trudged through a lush stand of old growth Devil's Club. We walked with the canoe upside down over our heads. I was in the lead since I was supposed to know the way. At some point we hit a nest of bald-faced hornets, which sped our progress considerably. Maybe we got lost for a while. It's hard to see anything with a canoe on your head. We made it back to the river, plastered our scrapes and stings with hypoallergenic mud and launched into the maelstrom. It was good to be alive.

July 20, 1998

Thanksgiving

You heard it here first. That's right, all the real news is in the back of this paper. This is going to be one long hard winter. Read the signs.

The old guide's woodpile is large enough to be photographed from space. The bucks on the upper Dungeness are as shaggy as German shepherds. And the spiders are flying as thick as I've ever seen.

Most folks don't know that spiders are good flyers but they are. Get up in the mountains on a calm, sunny, fall afternoon. Sit on a ridge top and look toward the sun. You'll see uncounted thousands of spiders flying, or para-sailing on diaphanous streamers of web. It looks like it's raining spun glass.

It's an odd feeling to be sitting in the high country dangling your feet into space and see a spider floating by. I'm not sure how they launch. But they do. Maybe they climb a tree, or jump off a cliff. Once spiders are airborne, they let out a big string of web and float where they will. They can traverse the Olympic Mountains in a day. I've seen spiders headed North for Canada. Some of them float to the east, toward the evil cities beyond the big water. Must be a shock for a poor spider to launch on the High Divide and land in Seattle. I know the feeling. I've never seen the spiders this thick. Look out.

The sandhill cranes came through in August. Sure they're just dumb birds. But they're going to Baja California for the winter. While I, man the tool user with the big brain, with gasoline, electricity and plastic, am going to stay here and try to survive one of the worst winters of the century. There are many ways to get ready for it. I suggest you max out that plastic and follow the cranes to Baja. Otherwise you'll be stuck here like me, and "what do the simple folk do?"

There are ancient rituals to endure hard times. I'll render a kettle of bear grease and do some baking. Wash it down with some green home brew. I picked a truckload of blueberries on the upper Sol Duc. I'll mash up some jam. Ward off the scurvy. Dig the potatoes. Run the smokehouse. Write the great American hillbilly novel about a simple fishing guide who starts a cult and...Don't let me ruin it for everyone.

I'm not going anywhere. I'll wait for the weather to hit. When it does we'll just let it. The rivers will flood but they'll drop down again. Rising and falling like the breath of a beast. We'll be there when the river turns from brown to green, hunting the winter steelhead.

So it's a little cold. Toughen up. There's nothing wrong with frozen fish. Ever heard of a heated drift boat? Whoever claims to have one is either a liar or a guide, or both. Forget the propane heaters when the ice pellets push upriver on a thirty-knot breeze. You're better off wrapped in R-40 insulation and black plastic. Hide under the splash deck with a thermos of hot buttered cocoa. Batten down the hatches. Enjoy the view.

Maybe you're allergic to frostbite. You might prefer to fish in the fall. The fish start to run with the rain. The rain quits and we'll have these perfect Indian summer days that I'll remember all winter.

The fish race upriver to spawn and die. We'll sit and watch them porpoise silver and blue through the tidewater. Later we'll see them ragged and black, washed up in the trees by the floods of winter. Life goes on.

Go ahead, make a cast. Doesn't matter what with. Get it in front of them. They'll either hit it or they won't. There's another run coming. Kings, silvers, and steelhead. We don't know what we'll catch until the end of the day.

We ride the white water through tunnels of red and gold leaves, under canyons of five hundred year old trees. If there was a more beautiful place on the planet I'd move there, but there ain't. I'm not going anywhere this winter. I don't want to miss anything.

Wouldn't you know it. It started raining in the time it took to write this. It's a whole new season. The harvest, time to appreciate what you've got. Thanks to all you folks who read this stuff. If I only believed half I heard, you people are nice. Maybe a little too nice.

I'd like to thank the folks who make this column so easy to write. My underhanded overbearing editor. The bought and paid for government do-gooders. The cut and run multi-national timber companies. The Makah tribe. But space does not allow.

September 30, 1998

"A Lone Gunman Shoots a Bus Driver and Himself, Plunging the Bus Off the I-50 Aurora Bridge."

Seattle Times, 12/19/98

Holiday Stress

It was another tough week in the news, although I would only know by accident. Up in the hills there's only Canadian television. The world could end this minute and they'd wait till the end of Canucks' hockey (pronounced Hawkee) highlights to give you a clue.

The Canadian language contains 50 different words for ice-skating, eh? On a good day I can get curling, hockey, and figure skating on the three different channels at the same time, which really gives the remote a workout.

Except for that yummy Australian soap opera, I would give up on TV entirely. Newspapers are useless. The only papers I get are old ones from New York City. I can tell you who's playing the Garden, the Met, or the clubs in the village, but I missed Jerry Miller right down here at the Carlsborg airport.

The phone is a total dud. The phone line is either getting strafed by nimrods who've run out of empty beer cans and want to shoot at a more challenging target, or ripped out by mudslides along the river.

One day a herd of elk got tangled up in the phone line. Don't worry, the elk are okay. The phone is out. I've been informed it's illegal to snare, trap or otherwise mess with an elk's day, so if it happens again I'll be in hot water.

This leaves my only source of news or current events the harried pilgrims from the concrete jungle who come to Lake Morning Wood Lodge for respite from the insanity that is the modern world. They are always nervous and in a hurry at first. Then they catch a fish and tell me what's the problem.

"Some guy shot a bus driver on the Aurora Bridge," my Seattle Metro bus driver client informed me. He described the flight of the bus off the high bridge, the landing on the apartment complex. You wonder what could make a guy (it's always a guy pulling stunts like this) do something that rotten.

"Holiday stress," the bus driver said. "I started getting flipped off two days before Thanksgiving. "Must escape city." He made another cast.

That's one of the perks of guiding. You learn about people from all walks of life. A Seattle bus driver is not allowed to wear a bulletproof vest or carry a gun. The drivers sit with their backs to a crowd of statistically hostile, deranged, self-medicating people while piloting a bloated cattle car through some of the worst traffic in the nation.

"I don't worry so much about the ones who say they're going to kill me or beat me up," the bus driver said. "It's the quiet ones I worry about." He accepts these shameful conditions with the sullen resignation that things will only get worse after the made-for-TV movie comes out. He uses humor to defuse violence.

"They can only shoot and kill me once, " the bus driver chuckled. "It's not like divorce."

No doubt about it. Seattle is a town on a bummer. Wild Billy (Gates) is getting whipped by the Black Queen. The Lazy B (Boeing) is laying off for the holidays, and the Sonics are on strike.

"We need more holidays," the bus driver says, "to take the pressure off the ones we have. It's winter. It's dark. People are depressed. Let's party."

He's right you know. The way we have holidays rigged now it's like the opening day of trout season or something. The holidays have become a carnival of greed, debt, and mindless consumption that determines the health of our nation's economy.

I say we don't mess with success. Let's celebrate more holidays, invent new ones. Stretch Christmas out 'til January. Do the Russian Orthodox Christmas. That's after the Boxing Day sales. Think of the money you'd save by doing Christmas later. There's a tip you won't hear from Martha Stewart, even though I worship her from afar.

Then there are Chinese New Year, Robbie Burns Day, and more holidays than I've ever heard of if you people would just start making a big deal about them. It's that simple. You want less holiday stress? Start making more holidays.

December 23, 1998

Christmas on a Budget

Lately someone asked about the fishing. That's a tough one. This hard rain falling week after dismal week doesn't mean you have to hang up the fishing pole. But when the big timber starts rolling down the river it's time to switch to the high water tactics.

The best way to get a fish once the river goes over the road is to run them over with a truck. Some folks claim this is illegal, but go ahead and check the fishing regulations.

It's a hundred and twenty pages of gibberish that says "No" to almost every form of fun there is; no blasting, no shooting, no poison. But not a word about running fish over with a truck. And now studded tires are legal!

As for me, I have been busy bird watching down the barrel of a shotgun. Once I shoot something, then I am duck hunting. Which brings me to the point of this article: What Christmas means to me. It's more than just a day off from duck hunting.

Christmas is a time off from our busy out-of-control hollow shell of existence life styles, to thank the good Lord we are not poor.

I know it's hard to believe, but I wasn't always a rich and famous award-winning writer.[2] Once I was so broke at Christmas I couldn't afford a dingleball for Lulu's Humvee. And if you think I let a little thing

---

2 Editor's note: Pat Neal still hasn't won an award for anything.

like being broke ruin Christmas, okay, you're right. What do you think? I'm running for Santa or something?

The point being I'm older and wiser now and I've learned it's okay to be broke at Christmas even though I'm not. All you really need is a little advance planning and a bunch of stuff you can probably borrow from a neighbor, if you have one.

So make a list and check it twice. You'll need some glue, a hammer, a bigger hammer, tin snips, fishing line, beer cans, hay wire, a shotgun, shotgun shells, a 55-gallon drum, a splatter shield, spray paint, glitter, and candles.The first rule of being broke at Christmas is to start your shopping early, which means if you are reading this now, you are already too late. You missed the big riots in Seattle. Too bad. Even if you were too gutless to do some holiday looting at the WTO riots, you might have picked up some unexploded concussion grenades or tear gas canisters. They make great stocking stuffers.

So what if you missed the riot. It's never too late for some holiday cheer. Just don't put all your projects off until Christmas Eve morning and you'll do fine.

Maybe you were so busy duck hunting you forgot all about Christmas. No problem. Empty shotgun shells make great tree ornaments. Thread them on fishing line. Spray them with paint, smear them with glue, sprinkle with glitter and you're done. I'll be giving one of these unique creations to the entire staff of the Sequim Gazette.

Next, take that empty case of shotgun shells. Spray it with the paint, smear with the glue, sprinkle with glitter. Then take some empty shotgun shells and give them the same treatment, only glue on some duck feathers. Now you have a manger scene complete with angels! I'm going to give one of these beauties to the guide association!

Now take an empty beer can. Blast it with a shotgun from about thirty yards away. This will blow a lot of little holes in it. Glue a candle in the bottom of the can, paint, glue, sprinkle, hang it up with the hay wire. You just made a Y2K light bulb! Everyone will want one in about a week or so.

Some sourpuss do-gooders claim Christmas is an environmental nightmare just because all the paper and ribbons clog up the recycling. There are alternatives.

Take the 55-gallon drum. Shoot some holes in it with the shotgun at close range for good air flow. For safety's sake remember your splatter shield. Spray paint the blasted barrel, glue, and glitter and there you have it! A holiday burning barrel. I'm making one for my editor. He wants to burn garbage when he grows up.

So what if it's getting late, you're out of shotgun shells and there's still someone on your list? No problem. Find another empty beer can somewhere. Smash it flat with a hammer. (Use the big one.) Hang the can on a wire. Cut it in the shape of a star with the tin snips. Paint, glue, glitter. Give it to…someone. I don't remember. The paint and glue fumes must be getting to me. Tidings of comfort and joy.

December 22, 1999

Robbie Burns Day

It was going to be one of those nights. The fog hung as thick as dragon's breath. A full moon lit the gloaming. Something was going to happen but I couldn't say what. I had my full curdling the haggis, which is not as easy as it sounds.

The lodge had been taken over by a flock of Robbie Burns fans celebrating the life and work of that 18th Century rake and poet. A failed farmer and inveterate fornicator, Burns began impregnating women at an early age. First a servant girl, then one Jean Amour who bore him twins. Her father was not amused. He broke up their common law marriage and went after Robbie for child support.

Robbie thought it wise to immigrate to Jamaica. He had a job offer as a slave driver-accountant on a large plantation. To raise money for the voyage and to express his own version of the goodbye cruel world theme, Burns published a collection of his poems in the Scots dialect.

Written in a state of abject poverty, defiance, and despair, the collection spared no one. The church and the devil, peasants and lairds were exposed to Burns's withering satire and comedic vision. He mined new depths of self-pity. He penned some of the most tender expressions of human affection known to exist.

Two hundred years later people still gather in the highlands to celebrate the work of one man who would not let the bastards get him down. Which explains why I was boiling the haggis.

Haggis is the national dish of Scotland. It is the sumptuous epitome of the phrase, "waste not, want not." It is a mixture of heart, lungs, kidneys, liver, suet, meat, oats, and spices sewn up in a deer's stomach and boiled to perfection.

Sure I had to make do with sheep parts, since all the deer hunters had thoughtlessly tossed their gut piles last fall. The only problem was to find a kettle big enough to give the thing a proper scald. Once I got the mess boiling, I thought I could relax, but no.

A demented shriek of pure terror rent the silence of the wilderness. One of the lodge inmates had been out in the meadow watching the moon. At first they thought they saw a shadow on the edge of the orb. Soon it appeared as if something was actually devouring the moon.

Of course, I'd known for some time that we were due for a lunar eclipse, but for whatever reason I'd neglected to inform the others. Knowledge is power. I thought it would be fun.

Eclipses are one of the oldest events recorded in human history. Four thousand years ago in China a pair of magicians had been entrusted with keeping the sun and the moon in a harmony that would prevent eclipses. They decided to get drunk instead. When an eclipse actually happened, the king sent a military expedition to punish the magicians.

Sure hindsight is 20-20. With the lens of history we can now say without a doubt this was just silly superstition. The Makah Indians of Cape Flattery have the only plausible explanation for an eclipse I've ever heard.

By now it is common knowledge the moon is composed of a jelly-like substance that is irresistible to the giant codfish, Toosh-Kow. So we watched with horror as the evil codfish slowly consumed the moon.

We had to do something. "Sacrifice a virgin!" someone suggested. That was no good. Human sacrifice has been politically incorrect since the days of the Aztecs. Besides, your chances of finding a virgin this far upriver are slim to none.

"Burn the witch!" the mob chanted. That was another no-brainer. The woods are mighty wet this time of year. It's tough to get a big fire going in a hurry when you really need one.

There was only one thing left to do. The Makah said if you made a loud noise it would scare the codfish away. I issued some war surplus Mexican seal bombs to the crew on the theory that it is better to light one seal bomb than to curse the darkness.

Just then I heard a loud explosion that blew the windows out of the bottom story of the lodge. I knew without looking what had happened. Like an idiot I'd forgotten to prick the haggis. With no release, the steam pressure had built up inside the sheep's stomach to the point of bursting, coating my designer kitchen from floor to ceiling with a slurry of half-cooked offal.

It was okay. The codfish was scared away. I turned on the fire hose and tidied up. I found some old hot dogs in my drift boat. And a happy Burns day to you.

January 19, 2000

Valentine's Day

A storm came to roost in the hills. It would not go away. The road washed out and clogged up with downed timber. Phone gone. Perfect. There was nothing to do but sit on the porch of the lodge and watch the roof leak. I would have gone up there and patched it, but you'd have to be nuts to work on a roof in the rain. And if it wasn't raining, I'd be fishing. There's just not enough hours in the day.

Then it started to snow. That froze up the roof and stopped the leak. Good thing. Lulu was having a cake party and you want to keep a close eye on kids these days to make sure you get your share.

I home-schooled my kids. There was only one classroom, three million acres of raw wilderness. There was only one lesson. The meaning of life. Then I sent them to public school anyway. What a dirty trick. I hope they never find out.

Lulu said when she grows up she will have a blue and white house with a cat and a dog. She will take care of me when I am old. So you can rat hole all the cash you want in some anonymous retirement thing. It does not love you. I have the real social security.

Of course this meant she'd take over my life. So what else was new? Starting with my socks didn't match; I had hat hair, and dead flowers on the table. To Lulu this meant I needed a girlfriend. Not a problem. She got on the web.

Meet Donna. She's one of Barbie's friends. Donna likes to meet people, talk to people and help people. She has the blonde haired, blue eyed, Nordic good looks of a Hitler Youth poster child, with sparkle accessories.

Do you believe in love at first sight? I'm certain that it happens all the time.

"Dad, she's not real." Lulu said. To you oldsters, that's pre-teen lingo for "Donna is a real doll." Whatever seal bombs life has tossed at our family, I've kept the communication open. I'm very proud of that.

Donna's so cute I'll bet she doesn't even go to the bathroom. Perfect. Maybe a little too perfect. But a guy can't be too choosy out here in the brush.

And just in time for my Valentine's Day column. It's a long shot sneaking this stuff into a paper with the editorial guidelines of Pravda in the old days. But it's worth a try. Anything for Donna.

So here goes. The ten most romantic places on the Olympic Peninsula. We'll start with something easy and close to home, and work our way out to where only kooks with waiver forms would dare to go. Warning, don't make people rescue you. They will not be happy. They will blame me.

1. Lake Morning Wood Lodge: Voted "Best Robbie Burns night on the Peninsula." Home of the Firbolg, and the "Curt Cobain Memorial Shotgun Classic." Enjoy a candlelit dinner in the biggest elk wallow on the Olympic Peninsula. Order the chicken fried duck and the blackberry fool, because that's all they serve. Dress: Warmly. Credit Cards: Do not let these people find out you have credit cards.

2. Blue Mountain snow caves: Hundred-mile views from the Cascades to the far end of Vancouver Island that will burn out your eyeball sockets at sunset. Too bad the government do-gooders blocked off the road. Why? Vandalism, snow, wrecks. Using this standard of maintenance they could shut down every road in the county. Why start with the best one we have? If you make it to the top of Blue Mountain these

days you will be in shape, and alone. Dress: Think Peary Expedition. Credit Cards: Great for identifying avalanche victims.

3. Cape Flattery: Most northwesterly point in the country. Best bird watching too. See sea otters, sea lions, seals and other vermin. Check out the Native American whaling culture. Watch some whales. While you can. Dress: Rain gear. Credit Cards: The universal language.

4. Kalaloch Big Cedar: Crawl up inside a giant cedar tree. Freud would have loved this one. But not on a windy day. Dress: Coveralls, ropes. Credit Cards: Keep your eyes open you might find someone's wallet.

5. Enchanted Valley: A.K.A. "valley of ten thousand waterfalls." These can freeze, thaw and flood in the same day trapping you in there. What could possibly be more romantic than that? Dress: Skis, snowshoes, chest waders. Credit Cards: Bring lots. Maybe you can use them to start a fire.

February 10, 1999

Two students armed with semi-automatic handguns, shotguns, and explosives conducted an assault on Columbine High School."

Seattle Times, 5/20/1999

Mothers Day

This was a bad week to read the paper. The news was bad and getting worse. It's a tough job trying to make sense of it. Is it the end of civilization, as we know it? Or just another millennial doomsday scam?

Don't get me wrong. I am all for the end of civilization as we know it. As long as I don't lose my health insurance.

I'm okay with the millennial doomsday thing too. I'm the only guide around with an end of the world fishing discount, or a nude fishing discount, too, but that's another column.

Old Bill Yeats had the best line about the "20 centuries" anyway. "Things fall apart, the center cannot hold." That's the great thing about history or poetry. You can just figure that all the really tough thinking in this world has already been done for you.

Besides, this is a Mother's Day column. Not all fishing guides were hatched under rocks. Some of us have moms. I have a great mom. I know because I am not writing this from prison.

Ma always encouraged us kids to pursue an active lifestyle. Maybe you think you have it tough driving your kids around to sports and stuff. I know I do. But at least we have good roads. It's nothing compared to what my poor mother, and my friends' moms had to drive.

That was some driving. They would drop us off at Deer Fly Park or Obstruction Point one day and pick us up at Dosewallips or Quinault a week later.

So when Mother's Day rolled around I always wanted to get her something cool. One day at the beach, with a gang of half-wit juvenile delinquent buddies who for some reason haven't been thrown in jail yet either, we found a dead sea lion.

We figured a fisherman plugged it. What a beautiful animal. It had a short golden fur. There was only one hole in it. It was almost Mother's Day. Suddenly I had a fantastic idea.

What mother wouldn't want a freshly tanned sea lion hide? It would make a great rug, or heck, a whole wall-to-wall carpet. There was nothing to do but get to work and skin it out.

I'd done some varmint hunting before, so I had some idea about skinning. I'd bagged a mountain beaver (it was him or me), and stashed the hide until the price of mountain beaver fur went up or the mice ate it. I don't remember.

At the time I thought there was a big future in varmint hunting. Coon hides were rumored to hit fifty bucks apiece any time. We skinned a few and tried to figure how to spend all the money we were going to make.

Skinning a sea lion couldn't be much different. Except they weigh 1,000 pounds. It was okay. There were four of us. A couple of us had pocketknives.

When I was a kid I was always very proud of having a sharp knife. That was because I always lost them before they had a chance to get dull. But even a sharp knife is no match for a sea lion hide. We rassled that carcass for what seemed like hours. It was one of the greasiest, filthiest chores you could tackle, but heck, it was almost Mother's Day.

We didn't exactly do a clean job of skinning. There were still a hundred pounds or so of blubber on the hide when we started dragging it down the beach. But it couldn't have weighed much more than two hundred and fifty pounds.

I remember wondering what we would have told a curious game warden if one had happened along. I'm sure we could have pleaded simple, brute dumbness and been exonerated of all charges. Later I wished we had seen a game warden. It would have saved a lot of trouble.

As it was we tossed the now muddy, sandy, greasy, bloody hide in the trunk of someone's car, or more precisely someone's mother's car, and headed for home.

It was a big job that took a lot of lumber to stretch that hide, but I did a good job of it. It must have been about ten feet square. A real trophy.

The sun came out. Molten blubber began running off in streams. A cloud of blowflies suddenly appeared. I moved the hide into the shade but it didn't seem to help. It turned black and rancid in an amazingly short time and dried overnight to the hardness of spring steel. The fur was still prime though. Ma said she liked it.

May 22, 1999

The Horse Whimperer

I've never caught a fish while digging fence postholes. Still, it's a great way to spend the weekend. Misery loves company, so get someone to help you. Tell them you're looking for gold. Have plenty of refreshments.

Thrust your shovel into the rich mellow earth. Hit a rock bigger than a breadbox. Pry at it a while with an iron bar, jackhammer, dynamite. Collapse sweating, bleeding, and whimpering in the shade. Rise up and dig another, and another, on into the horizon.

Then stuff some posts or other into the holes and you're ready to call it a day. But you can't. The posts don't work for a fence by themselves. Many others and I have tried the honor system but it's just another failed attempt to reason with dumb brutes. You have to string some wire or nail something between the posts or it just won't work.

When in the course of human events you should happen to finish the fence, you can begin the next phase of the operation, which is fixing the fence.

Because, while you're busy healing up your blisters and visiting the chiropractor, the barnyard vermin have spent every idle moment, and they have plenty, scamming the inevitable breakout.

I used to take this personally. Watching my finely crafted handiwork being used as a scratching post by a large hairy rump no matter how many times I've told my editor he should get that rash checked out.

Then I grew to accept my station in life as a career fence builder. Until the day I became an animal rights activist. It happened one night when I had enough gasoline to run the television.

An animal rights spokesperson who just happened to be a yummy soap opera star, said zoos and circuses should be outlawed because they restrict an animal's freedom, and freedom is what it's all about.

I really had no right to fence in another sensitive intelligent life form and deny its pursuit of happiness. I think it's in the Constitution, or should be.

I tried to explain the simple logic of this beautiful idea to Lulu when she moved her pack string onto the summer ground. I don't like to brag about it but I'm a pretty darned good parent. I buy my kids whatever they could possibly want and do whatever they say in hopes they won't eventually shoot me. Who knows, they might even talk to me when they grow up, or even loan me money.

Still, the animal rights thing did not fly with this cowgirl. After showing me where I could stick the next fence line she told me to saddle up this big mean looking nag and not mess it up this time.

This was yet another outrage to my animal rights sensitivities. Riding horses is a cruel form of animal slavery. It is an outmoded abuse of another intelligent life form. This too went over like a rock in a feedbag with Lulu. I was going to ride that horse and that was that.

Like most people, I've learned everything I know about horses by watching TV and movies. The secret is to whisper to the horse. Tell it how much you love it and it will love you and maybe not scrape you off on a tree up in the brush somewhere, leaving you something for the ravens to find.

The last time I tried whispering to a horse about how much I loved her, she stepped on my foot. I'm okay now. It's amazing what they can do these days with prosthetics. So I gave up on the love thing. Now I just beg them not to hurt me. Call me the horse whimperer.

I couldn't find a ladder so I lead the beast to a high stump and crawled up on the monster's back. I was sure I could feel the love. Then the little ears started twitching and the tail started swishing, which is horsy talk for you'd better "choke the biscuit". That means hang on to the saddle horn. Some people claim they don't. I do, for dear life.

"He likes to jump!" Lulu yelled. It was just another case of "now she tells me."

There's no good way to get off a horse when it's running flat out on a mountain trail, jumping logs and ditches.

It was like being strapped on to a big hairy rocket. We made it to the top of the ridge. A yellow moon rose through a purple sky. I wasn't bucked off yet. It was good to be alive.

May 26, 1999

"Governor Locke Visits Port Angeles Declaring it State Capital For a Day"

Sequim Gazette, 11/24/99

False Dungeness

It was another tough week in the news. The governor came to town. He made False Dungeness the capital of the State of Washington for a day. It seemed like a good idea at the time. I thought the governor or one of his fancy friends might book a trip to do their part to help the state employment numbers by hiring a humble fishing guide, but they didn't, so here goes.

Sure you Jimmie-come-latelys call it Port Angeles. But that town has had more names than my ex-wife. Old Captain Eliza called it "Puerto de Nuestro Senora de los Angeles," when he anchored up in 1795.

That's a pretty good name I'd say. They should have stuck with it. Think of the name recognition spin-off cottage industries that would have spawned by calling Port Angeles, Los Angeles instead.

Doesn't matter how we get the tourists here. It's okay if they're lost. It just costs more to get home. Shanghaiing was a proud pioneer tradition here on the Olympic Peninsula. With your sailors always jumping ship to go live with the Indians, what was an old time sea captain to do but recruit a crew from the local crop of farm boys?

"We must look to our past to grasp our future," the governor said as he wolfed down his free spaghetti dinner at the community hall. Okay, maybe he didn't say that, but he should have.

I wouldn't know, but I have a very bad feeling about letting that P.A. bunch imagine they are the capital of anything.

It was called "False Dungeness" by generations of lost sea captains, who being confused by the similar appearance of the Dungeness and Ediz sand spits dropped their pick in the wrong place.

This was a mistake. False Dungeness had three Indian villages sure, but their inhabitants may have been decimated by the white man's disease even before the white man historically "arrived." So False Dungeness was uninhabited much of the time, which was no fun for lonely sailors.

The real party was down the beach at New Dungeness, which was also known as Whiskey Flats. Confused? Hang on, it gets a lot weirder.

Old Captain Vancouver must have been mighty homesick when he named it New Dungeness back in the spring of 1792. Spring has always been the best time to sell real estate. The flowers were blooming, the grass was green. No telling what Captain Vancouver would have called it later in the summer when everything but the cactus dried up and blew away.

It didn't matter to a rabble of Americans who showed up and named the place Whiskey Flats, commemorating the town's first industry, selling liquor to the Indians.

Business was good for a while. Whiskey Flats became our first county seat. But even way back then Dungeness Bay was filling up with silt. This was before there were any loggers to blame.

So they moved the town east across the river and called it Dungeness. Whereby New Dungeness became Old Dungeness. Meanwhile False Dungeness languished for a time as "Windsor's Harbor," a name given by old Judge Swan from the deck of his party boat. I think it was the rum talking.

Victor Smith, founding father of Port Angeles and patron saint of real estate speculators, wanted to name it "New Cherbourg." That didn't stick either. It was okay. Smith is still a hero in P.A. He personally stole the customs house at gunpoint from Port Townsend and set it up in Port Angeles. He got President Abraham Lincoln to declare Port Angeles a Federal Reserve and "Second National City." That was so if anything happened to Washington, D.C., we'd have another capital all ready and waiting, three thousand miles away.

This made a lot of sense back then. Especially if you happened to own land in Port Angeles, like Mr. Smith and his cronies.

Port Angeles has always been a hotbed of megalomaniacal community boosters. Just ask Whiskey Flats AKA New-Old Dungeness. By 1890 it had fallen on hard times. Once the Indians died off, the whiskey trade dried up for some reason.

Port Angeles used an illegal election and a gang of vigilantes to steal the county seat from Whiskey Flats, beating Port Crescent to the punch.

Now Port Angeles wants to be the state capital. Sure they say it's only for a day, but that's one day too long for us old timers back in Whiskey Flats. As for Governor Locke, I can only say, those who ignore history are doomed to watch television.

December 1, 1999

A Hundred Years from Nowhere

It was twilight in the meadow where the old cabin stood. One look told me I was more than a little late for dinner. It was okay. No one had been home for dinner since the '30s. The door had been chopped up for kindling in the '70s. The roof caved in during that big snow in '85. This was a homestead cabin.

The Homestead Act of 1862 was the coolest thing to ever hit this country, unless you were an Indian. They were already living in all of the best places, the mouths of streams, harbors, and prairies. This was a problem until the Homestead Act.

Old Dan Pullen burned the Quileute village when he staked his claim in LaPush. The Quileutes were in Puyallup picking hops at the time. Dan probably thought they'd moved. Sure. This is what historians call a "pattern of settlement."

All you had to do was build a cabin at least twelve feet square. Do some cultivation. Live there five years. You've got clear title to 160 acres of prime bottomland.

What if you were busy fishing, or too lazy and shiftless to build a cabin or raise a garden or hang around for five years? Not a problem. Stack up a few logs. Set a potted plant on a stump. That's a residence and cultivation, if anyone asks.

Pressed for time? Buy your claim outright for a buck and a quarter an acre. Need cash now? Sell out to a friendly neighborhood timber company. They'll log it and let the land go back to the government for taxes, like they'd just "borrowed" it.

Meanwhile you, the rugged Homesteader, just made yourself fifty bucks. That was back when fifty bucks was worth something. The government really was here to help you, for a fee.

The more homestead claims you filed, the easier it got. It was a beautiful system until some knee-jerk do-gooder started whining about "the looting of the public domain," and locked up what was left as a Forest Reserve we couldn't log.

It was okay. Enough people had already been lured here with flowery promises of strawberries picked in December, roses in January. From the harbor that could "accommodate all the fleets of the world," to the mountains "filled with coal and precious metals," there was no frost in the ground in winter, no heat or bugs in summer. Potatoes grew the size of rutabagas, rutabagas the size of melons. The mountains, creeks, and lakes were named after all the iron, gold, and silver you were sure to find there. Promotions like this put Copper City on the map. All that we needed was a little hard work and risk capital to become the next Butte, Montana!

Imagine the shock of the Jimmie-come-latelys. Like the pioneer woman hauled ashore on the back of an Indian. There's symbolism for you. The prime bottomland was already taken by squatters, who stole it from the Indians, just as soon as it was legal. The come-lately people had to go upriver.

That story about there being a bunchgrass prairie and a big lake up in those mountains was just that, a story. There's nothing up there but more mountains. The highest one upriver being named Mt. Deception. Just a coincidence I'm sure.

Did we mention the forest fire? Burnt up most of the timber, so don't plan on selling out in your lifetime. Find a flat on a side hill. Water would be nice. A garden required. Better plant those taters and rutabagas quick. It could be all you and your stock have to live on all winter.

Welcome to Subsistence Street. A rich man might have had a dairy cow. But with no refrigeration or transportation he fed the milk to the hog, if he had one. He could butcher the hog, cure some hams, paddle them twenty miles across the treacherous Strait of Juan de Fuca and sell them in Victoria. Predictably, no one ever got rich off this convoluted food chain. Maybe you could earn some cash trapping...good luck. The Hudson's Bay Company just got done trapping here before they moved out. Maybe Pa could leave and go logging in the camps while Ma stayed home and made sure the root cellar was plugged for the winter. Such was life on the mountain homestead.

By 1900, the only ones left were too broke to leave. They held on until the Depression when farming got a whole lot tougher. There's not much left of this split cedar culture now. Just a few old shacks rotting into the brush until they're finally gone, abandoned as the day they were found.

I like these pioneer home tours, particularly on a moonlit night when the wreckage of a lifetime's work reflects a ghostly light. I feel a hundred years from nowhere. It's the end of the last frontier.

November 11, 1998

9 780595 168125